Identifying and Reducing Environmental Health Risks of Chemicals in Our Society

WORKSHOP SUMMARY

Robert Pool and Erin Rusch, *Rapporteurs*

Roundtable on Environmental Health Sciences, Research, and Medicine

Board on Population Health and Public Health Practice

INSTITUTE OF MEDICINE
OF THE NATIONAL ACADEMIES

THE NATIONAL ACADEMIES PRESS
Washington, D.C.
www.nap.edu

THE NATIONAL ACADEMIES PRESS   500 Fifth Street, NW Washington, DC 20001

NOTICE: The workshop that is the subject of this workshop summary was approved by the Governing Board of the National Research Council, whose members are drawn from the councils of the National Academy of Sciences, the National Academy of Engineering, and the Institute of Medicine.

This activity was supported by contracts between the National Academy of Sciences and the National Institute of Environmental Health Sciences (HHSN26300033), The Kresge Foundation, Colgate-Palmolive Company, ExxonMobil Foundation, and Royal Dutch Shell. The views presented in this publication do not necessarily reflect the views of the organizations or agencies that provided support for the activity.

This summary is based on the proceedings of a workshop that was sponsored by the Roundtable on Environmental Health Sciences, Research, and Medicine. It is prepared in the form of a workshop summary by and in the name of the rapporteurs as an individually authored document.

International Standard Book Number-13: 978-0-309-30115-2
International Standard Book Number-10: 0-309-30115-7

Additional copies of this report are available from the National Academies Press, 500 Fifth Street, NW, Keck 360, Washington, DC 20001; (800) 624-6242 or (202) 334-3313; http://www.nap.edu.

For more information about the Institute of Medicine, visit the IOM home page at: **www.iom.edu.**

Printed and bound in Great Britain by Marston Book Services Ltd, Oxfordshire

The serpent has been a symbol of long life, healing, and knowledge among almost all cultures and religions since the beginning of recorded history. The serpent adopted as a logotype by the Institute of Medicine is a relief carving from ancient Greece, now held by the Staatliche Museen in Berlin.

Suggested citation: IOM (Institute of Medicine). 2014. *Identifying and reducing environmental health risks of chemicals in our society: Workshop summary.* Washington, DC: The National Academies Press.

*"Knowing is not enough; we must apply.
Willing is not enough; we must do."*
—Goethe

INSTITUTE OF MEDICINE

OF THE NATIONAL ACADEMIES

Advising the Nation. Improving Health.

THE NATIONAL ACADEMIES
Advisers to the Nation on Science, Engineering, and Medicine

The **National Academy of Sciences** is a private, nonprofit, self-perpetuating society of distinguished scholars engaged in scientific and engineering research, dedicated to the furtherance of science and technology and to their use for the general welfare. Upon the authority of the charter granted to it by the Congress in 1863, the Academy has a mandate that requires it to advise the federal government on scientific and technical matters. Dr. Ralph J. Cicerone is president of the National Academy of Sciences.

The **National Academy of Engineering** was established in 1964, under the charter of the National Academy of Sciences, as a parallel organization of outstanding engineers. It is autonomous in its administration and in the selection of its members, sharing with the National Academy of Sciences the responsibility for advising the federal government. The National Academy of Engineering also sponsors engineering programs aimed at meeting national needs, encourages education and research, and recognizes the superior achievements of engineers. Dr. C. D. Mote, Jr., is president of the National Academy of Engineering.

The **Institute of Medicine** was established in 1970 by the National Academy of Sciences to secure the services of eminent members of appropriate professions in the examination of policy matters pertaining to the health of the public. The Institute acts under the responsibility given to the National Academy of Sciences by its congressional charter to be an adviser to the federal government and, upon its own initiative, to identify issues of medical care, research, and education. Dr. Harvey V. Fineberg is president of the Institute of Medicine.

The **National Research Council** was organized by the National Academy of Sciences in 1916 to associate the broad community of science and technology with the Academy's purposes of furthering knowledge and advising the federal government. Functioning in accordance with general policies determined by the Academy, the Council has become the principal operating agency of both the National Academy of Sciences and the National Academy of Engineering in providing services to the government, the public, and the scientific and engineering communities. The Council is administered jointly by both Academies and the Institute of Medicine. Dr. Ralph J. Cicerone and Dr. C. D. Mote, Jr., are chair and vice chair, respectively, of the National Research Council.

www.national-academies.org

.

PLANNING COMMITTEE FOR THE WORKSHOP ON IDENTIFYING AND REDUCING ENVIRONMENTAL HEALTH RISKS OF CHEMICALS IN OUR SOCIETY[1]

DENNIS J. DEVLIN, ExxonMobil Corporation, Irving, TX

LYNN R. GOLDMAN, George Washington University, Washington, DC

WILLIAM E. HALPERIN, Rutgers New Jersey Medical School, Newark, NJ

AL McGARTLAND, U.S. Environmental Protection Agency, Washington, DC

SUSAN L. SANTOS, Rutgers School of Public Health, Piscataway, NJ

KIMBERLY THIGPEN TART, National Institute of Environmental Health Sciences, Research Triangle Park, NC

PATRICIA VERDUIN, Colgate-Palmolive Company, Piscataway, NJ

HAROLD ZENICK, U.S. Environmental Protection Agency, Research Triangle Park, NC

---

[1] Institute of Medicine planning committees are solely responsible for organizing the workshop, identifying topics, and choosing speakers. The responsibility for the published workshop summary rests with the workshop rapporteurs and the institution.

ROUNDTABLE ON ENVIRONMENTAL HEALTH SCIENCES, RESEARCH, AND MEDICINE[1]

---

[1] Institute of Medicine forums and roundtables do not issue, review, or approve individual documents. The responsibility for the published workshop summary rests with the workshop rapporteurs and the institution.

PAUL SANDIFER, National Oceanic and Atmospheric Administration, Charleston, SC
SUSAN L. SANTOS, Rutgers School of Public Health, Piscataway, NJ
JOHN D. SPENGLER, Harvard School of Public Health, Boston, MA
G. DAVID TILMAN, University of Minnesota, St. Paul
PATRICIA VERDUIN, Colgate-Palmolive Company, Piscataway, NJ
NSEDU OBOT WITHERSPOON, Children's Environmental Health Network, Washington, DC
HAROLD ZENICK, U.S. Environmental Protection Agency, Research Triangle Park, NC

IOM Staff

KATHLEEN STRATTON, Study Director (*from September 2013*)
ERIN RUSCH, Associate Program Officer
HOPE HARE, Administrative Assistant
ROSE MARIE MARTINEZ, Director, Board on Population Health and Public Health Practice

Reviewers

This workshop summary has been reviewed in draft form by individuals chosen for their diverse perspectives and technical expertise, in accordance with procedures approved by the National Research Council's Report Review Committee. The purpose of this independent review is to provide candid and critical comments that will assist the institution in making its published workshop summary as sound as possible and to ensure that the workshop summary meets institutional standards for objectivity, evidence, and responsiveness to the study charge. The review comments and draft manuscript remain confidential to protect the integrity of the process. We wish to thank the following individuals for their review of this workshop summary:

Johanna T. Dwyer, Tufts Medical Center
Jay Lemery, University of Colorado School of Medicine
Patricia Verduin, Colgate-Palmolive Company
Lauren Zeise, California Environmental Protection Agency

Although the reviewers listed above have provided many constructive comments and suggestions, they did not see the final draft of the workshop summary before its release. The review of this workshop summary was overseen by **Mark R. Cullen,** Stanford University. Appointed by the Institute of Medicine, he was responsible for making certain that an independent examination of this workshop summary was carried out in accordance with institutional procedures and that all review comments were carefully considered. Responsibility for the final content of this workshop summary rests entirely with the rapporteurs and the institution.

Contents

1

Introduction[1]

On November 7–8, 2013, the Institute of Medicine's Roundtable on Environmental Health Sciences, Research, and Medicine held a workshop to discuss approaches related to identifying and reducing potential environmental public health risks to new and existing industrial chemicals present in society. Industrial chemicals include chemicals used in industrial processes or commercial products, not including those found in food, pesticides, or pharmaceuticals. Through presentations and discussions, the workshop examined successes and areas for improvement within current regulatory programs for assessing industrial chemical safety, frameworks for chemical prioritization to inform targeted testing and risk management strategies, concepts of sustainability and green chemistry that support the design and use of safer alternatives, and efforts to reduce the risk of chemicals in our society. The workshop statement of task is provided in Box 1-1.

The following is a summary and synthesis of the presentations and discussions that took place during the 2 days of the workshop. The planning committee worked to identify varied perspectives on the topic areas included in the workshop, and the diversity of speakers and stated viewpoints contributed to a wide range of input on the subject of reducing the risk of chemicals in society. When reading the summary it is important to keep in mind that the opinions expressed and any recommendations made are those of the individual speakers themselves and do not represent the position of the Institute of Medicine or the National Academies. Indeed, the purpose of the Roundtable on Environ-

---

[1] The planning committee's role was limited to planning the workshop, and the workshop summary has been prepared by the workshop rapporteurs as a factual summary of what occurred at the workshop. Statements, recommendations, and opinions expressed are those of individual presenters and participants, and are not necessarily endorsed or verified by the Institute of Medicine, and they should not be construed as reflecting any group consensus.

BOX 1-1
Statement of Task

An ad hoc committee will plan and conduct a 2-day, public workshop on identifying and reducing environmental health risks of chemicals. The workshop will focus on responsibilities and authorities for safeguarding the public from chemical hazards at the federal, state or local, and global levels (e.g., Toxic Substances Control Act [TSCA], Strategic Approach to International Chemicals Management). Furthermore, invited speakers will present and discuss the process for assessing chemicals (e.g., prioritization of chemicals under TSCA) and the state of the science in chemical hazard assessment (adequacy of current models, innovative approaches), public health goals for managing risks of chemicals, protections for vulnerable populations, and communication with consumers. The committee will develop the workshop agenda, select invited speakers and discussants, and moderate the discussions. A workshop summary will be prepared by a designated rapporteur in accordance with National Research Council policies and procedures.

---

mental Health Sciences, Research, and Medicine is to provide a mechanism for interested parties in environmental health to meet and discuss sensitive and difficult environmental issues in a neutral setting. The Roundtable fosters rigorous dialogue about these issues, but it does not provide recommendations or even try to find a consensus on these issues.

ORGANIZATION OF THE SUMMARY

The organization of this summary roughly parallels the structure of the workshop itself with remarks by a few individuals regrouped to reflect the themes that developed over the 2 days. Chapter 2 begins with a summary of the challenge of chemicals in today's society and general approaches to dealing with chemical risk. Chapter 3 summarizes the presentations and discussion of the current regulatory approaches to dealing with industrial chemicals. Chapter 4 is devoted to providing background and context to the topic of chemical risks and models for environmental risk assessment. Chapter 5 summarizes presentations on improved approaches to priority setting in the risk assessment and risk management of industrial chemicals. Chapter 6 includes a summary of a variety of approaches that institutions have taken to reducing chemical risks in our society. Chapter 7 summarizes the final panel of the workshop, where various speakers synthesized and expanded on the presentations and discussions that had taken place on the previous day and a half.

KEY THEMES

Over the course of the workshop, several prominent themes emerged from the individual speakers' remarks and discussion sessions that ensued. These themes are presented here as a way to organize the material summarized in this report and do not represent conclusions or recommendations from the workshop.

- Individual workshop speakers noted that the volume of chemicals in commerce increased greatly during the 20th century and that government agencies continue to struggle to identify which chemicals are in use today in order to better understand hazards chemicals may pose to human health and the environment.
- Much of the discussion focused on the Toxic Substances Control Act (TSCA) of 1976,[2] with individual workshop participants agreeing that a legislative update could greatly improve the regulatory process for industrial chemicals in the United States.
- Although workshop speakers presented varying prioritization approaches to chemical testing and management, the matrices were quite similar in their characterization of hazard and exposure. The matrices tended to differ in the level of precaution employed, with some frameworks moving ahead boldly and others conservatively.
- It is clear that there are many drivers for reducing the risk of chemicals in our society (e.g., reducing costs, reducing risks, and such intangibles as the reputation of products and companies). Individual workshop participants noted that the life-cycle approach and improved availability of chemical information and assessment are important facets to achieving this goal.

---

[2] Toxic Substances Control Act of 1976, Public Law 94-469, 94th Congress.

2

The Challenge: Chemicals in Today's Society

To set the stage for the rest of the workshop, several presenters spoke about the challenge of chemicals in today's society and general approaches to dealing with chemical risk.

Lynn R. Goldman, Dean of the George Washington University School of Public Health and Former Assistant Administrator for Toxic Substances at the U.S. Environmental Protection Agency (EPA), began by offering some historical context. Twenty years ago she had joined the EPA, where she was responsible for the chemical and pesticide regulatory programs and the concerns then were very similar to those today. "Indeed, all of these topics that we have before us today were topics that were before the EPA at that time." She noted that chemicals regulation is a very difficult and complex area but for years it has been clear that the EPA has been unable to make adequate progress under existing requirements.

A fundamental notion in dealing with chemical hazards is risk, a concept that has been promulgated extensively by the National Academies, beginning with the 1983 publication *Risk Assessment in the Federal Government: Managing the Process* (NRC, 1983), known as the Red Book. The Red Book describes four steps for risk assessment: (1) hazard identification, (2) dose–response assessment, (3) exposure assessment, and (4) risk characterization. Goldman expanded on two of these components, hazard and exposure. Hazardousness is the ability of a chemical to actually cause harm at various dosage levels, Goldman said, while exposure is the amount of dose that might be received at target tissue after contact. Exposure may depend on various susceptibility factors such as age and stage of development, gender, genetics, nutrition, and comorbidities. "There are many individual issues that can cause variability in responses to chemicals," she said. "That means, of course, that the availability of scientific information is fundamental to our ability to understand risk, and it is also fundamental to our ability to manage those risks."

THE CHALLENGE

A good place to start in understanding the challenge facing the country is to get a sense of just how many chemicals are produced and used in society. To offer some historical context, Goldman quoted Paracelsus, the 16th-century Swiss-German physician, botanist, and alchemist who is credited with founding the science of toxicology. Listing the chemicals present in commerce nearly 500 years ago, Paracelsus wrote:

> What, then, shall we say about the receipts of Alchemy, and about the diversity of its vessels and instruments? These are furnaces, glasses, jars, waters, oils, limes, sulphurs, salts, salt-petres, alums, vitriols, chrysocollae, copper-greens, atraments, auri-pigments, fel vitri, ceruse, red earth, thucia, wax, lutum sapientiae, pounded glass, verdigris, soot, crocus of Mars, soap, crystal, arsenic, antimony, minium, elixir, lazarium, gold-leaf, salt-nitre, sal ammoniac, calamine stone, magnesia, bolus armenus, and many other things. Moreover, concerning preparations, putrefactions, digestions, probations, solutions, cementings, filtrations, reverberations, calcinations, graduations, rectifications, amalgamations, purgations, etc., with these alchemical books are crammed. Then, again, concerning herbs, roots, seeds, woods, stones, animals worms, bone dust, snail shells, other shells, and pitch. (Paracelsus, 1531)

"It was a fairly short list of chemicals," Goldman noted. "They lived in a world where most human needs, material needs, were met by the natural world through wood, metals, and other resources that were extracted from the natural environment. Today, we live in a very different world, where nearly everything in this room is in some way derived from industrial chemicals."

The volume of chemicals in commerce increased a great deal during the 20th century, she noted. In just the 25 years between 1970 and 1995, the volume of synthetic organic chemicals produced tripled, from about 50 million tons to approximately 150 million tons (see Figure 2-1) (Goldman, 2002). And today it is much more, she noted.

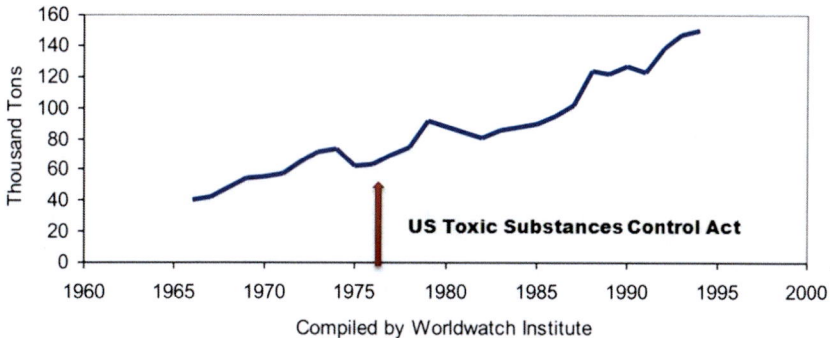

FIGURE 2-1 Volume of chemicals in commerce: U.S. synthetic organic chemical production, 1966–1994.
SOURCE: Adapted from Goldman, 2002. Copyright © 2002, Environmental Law Institute®, Washington, DC. Reprinted with permission.

Richard Denison, Senior Scientist at the Environmental Defense Fund, also spoke about the increase in chemical use. "Clearly, there has been a strong upward trend in just the sheer volume of chemicals being produced," he said. According to one estimate, global sale of chemicals has increased by a factor of about 25 since 1970, from $171 billion to $4.1 trillion (UNEP, 2013). Over the next few decades, he added, the rate of increase in the volume of chemicals used worldwide is expected to continue, or even accelerate (UNEP, 2013), and it will dramatically outpace the increase in population (see Figure 2-2) (Wilson et al., 2008).

The other thing that has changed, Denison said, is the diversity of use of chemicals, especially in consumer products and building materials, with manufactured substances replacing natural materials. One estimate is that chemicals are used in 96 percent of manufactured materials and products (ACC, 2014).

The Total Number of Chemicals in Commerce Today

Goldman noted that given the huge increase in the volume of chemicals produced in the United States and worldwide, an important question to ask is which chemicals are in use and how many are there. This is an important starting point for understanding the hazards that chemicals might pose to the environment and human health. It turns out, however, that this is not an easy question to answer.

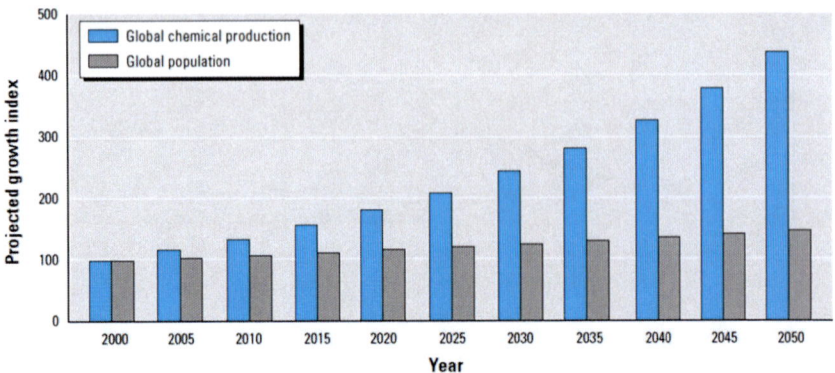

FIGURE 2-2 Projected growth in worldwide chemical production and global population through 2050.
SOURCE: Wilson and Schwarzman, 2009. Reprinted with permission.

Goldman explained that the best available estimate comes from the EPA, which was tasked by the Toxic Substances Control Act (TSCA) of 1976[1] with creating an inventory of chemicals being produced in this country. TSCA did not cover food and food additives, drugs, or cosmetics (all of which are regulated by the Food and Drug Administration), nor does it cover firearms and ammunition, pesticides, tobacco, and "mixtures," although the components of those things are covered. Otherwise, TSCA covers pretty much all of the chemical substances produced in the United States.

To create the initial chemicals inventory, Goldman said, TSCA required all manufacturers and processors of chemicals to report about those chemicals to the EPA between 1978 and 1982. During that period, the EPA received reports on approximately 62,000 chemicals (EPA, 2014).

The law also required that, when manufacturing a new substance that was not on the list, the manufacturer had to bring it to the EPA for review and to add it to the inventory using a process established in TSCA called Premanufacture Notification (PMN). Chemicals added via the PMN process are called "new chemicals." From 1982 and 2012, the agency added 22,000 new chemicals to the inventory. Therefore, the EPA inventory

---

[1] Toxic Substances Control Act of 1976, Public Law 94-469, 94th Congress.

now contains around 84,000 chemical substances that may possibly be in commerce (GAO, 2013).

TSCA also established a process for the EPA to periodically obtain updates on the manufacture, import, and use of chemicals on the inventory; the EPA obtains these data via regulation. Initially, the EPA updated the inventory every 4 years—in 1986, 1990, 1994, 1998, 2002, and 2006. Those updates were not complete, Goldman said. First, small manufacturers were exempted from reporting if they were manufacturing less than 10,000 pounds of a substance at a single site during the time of a report. Second, inorganics, polymers, microorganisms, and naturally occurring chemicals were exempted. In 2006, exemptions were expanded when the exemption cutoff for small manufacturers was increased to 25,000 pounds per year at a single site at the time of the report, and petroleum process streams and certain forms of natural gas were exempted from reporting. At the same time, in 2006 some data collection was expanded through the phase-in of reporting of inorganics and new requirements to report on the use of chemicals and production data for the chemicals with the highest production volumes. In 2012 the EPA revised the regulation yet again, gradually lowering the volume thresholds for reporting to increase the total number of chemicals that must be reported.

"In reality," Goldman said, "we really don't know how many chemicals are currently in commerce in the United States because the method of updating the inventory wasn't designed to answer that question." Instead, it was designed to inform us about chemicals that are produced at higher volumes.

The maximum possible number is about 84,000 chemicals in commerce, she said, but many new chemicals actually never make it to the market even though they were put on the inventory. A company may choose not to bring a chemical to market for many reasons, and it does not report that to the EPA. Furthermore, there is no process for delisting existing chemicals that are no longer in commerce. The Society of Chemical Manufacturers and Affiliates reports that there are about 25,000 chemicals in commerce (SOCMA, 2014), but this is probably a minimum estimate, Goldman said. So there are somewhere between 25,000 and 84,000 chemicals in commerce in the United States. "That is a pretty wide range," Goldman commented, and the uncertainty—in both the number of the chemicals and even the identities of which are in circulation—offers an indication of the problems that arise in trying to prioritize the various chemicals.

The Changing Understanding of Chemical Exposures

In addition to the growing number and volume of chemicals produced by society over the past several decades, Denison said, a second major change has been in our knowledge about how chemical exposures may occur. "The advent of biomonitoring during this period has shown that we all carry around hundreds of synthetic chemicals in our bodies," he said. "Every time we look for more, we find more." That realization has been combined with a growing understanding of how chemicals move though the environment—via both air and water and sometimes over quite long distances—and how chemicals that are used in products may make their way into human bodies.

Denison offered two examples of how such movement from products into the environment and into people can occur. In recent years, he said, researchers have determined how the flame retardants used in furniture foam end up within people. "Every time you sit on an upholstered item, a little bit of dust puffs out," he said, "and that dust includes those chemicals." The dust can be either ingested or inhaled. "That is a pretty clearly established pathway for chemical exposure that we didn't ever really think about, and certainly not several decades ago."

The second example involves coal tar–based sealants used on parking lots. The U.S. Geological Survey has tracked the sources of polycyclic aromatic hydrocarbons (PAHs) found in urban sediments and streams and discovered that a major source is contributed by runoff from parking lots treated with such sealants. The researchers have now extended that work, Denison said, and found that "in apartment buildings adjacent to parking lots that are treated with these sealants, people are literally tracking this material into their homes, and it is resulting in higher levels of PAHs in the house dust in those homes."

There has also been a growing realization in recent years that chemical exposures often affect different populations disproportionately—and that it is often those in lower socioeconomic brackets who suffer most. "That raises a lot of environmental justice concerns that we are much more cognizant of today," he said.

There have been various drivers for the growing concern over chemical exposures, Denison said. Medical science has shown, for example, that a number of specific chronic diseases are on the rise in the human population despite an overall trend of reduction in chronic disease. For instance, Denison noted that childhood cancers and leukemia are becoming more common (Ward et al., 2014), as are infertility and other reproductive problems and learning and developmental disabilities (Safer Chemicals Healthy Families, 2012).

Certain chemicals are being linked to these same chronic diseases, both from studies in laboratory animals and sometimes also from epidemiologic data. "Now, that is still a circumstantial case in many cases," Denison said, "but it is increasingly one that is showing connections between those exposures and diseases and disorders that are rising in the human population faster than genetics or something like that could explain."

This in turn has led to a growing recognition of the various ways people may be susceptible to chemical exposures. Researchers now realize, for instance, that early-life exposures can have very significant effects, some of which can last a lifetime. And exposures to chemicals that mimic biologically active chemicals that are normally found in our bodies, such as hormones, can exert effects, especially early in development, and even at low doses. And there has been a growing understanding of how epigenetics may be a mediator for chemical and other environmental exposures that may also help explain some of the variability in susceptibility that has been observed. Epigenetics offers a basis for understanding how early-life exposures can lead to later-life health repercussions, including different disorders and diseases. It is even possible that epigenetics could lead to transgenerational effects, Denison said, although that is still a very controversial concept.

The scientific approach to assessing risks has also changed dramatically in the past few decades, Denison said, particularly in the period between the 1983 publication of *Risk Assessment in the Federal Government: Managing the Process*, known as the Red Book (NRC, 1983), and the 2009 publication of *Science and Decisions: Advancing Risk Assessment* (NRC, 2009). "There are a lot of issues that are on the table now," he said, "and part of that debate is how to better assess and take into account human variability." Researchers are increasingly recognizing the presence of such variability in the human population, both from genetic variations and from other differences, such as variations in nutrition, health, and ways of living.

Denison explained that this variability raises questions about how one deals with the uncertainty associated with identifying a level of concern for a chemical. How much variability is there, for example, in the dose response to various chemicals? Does it still make sense to consider that there are thresholds below which there is no effect? Even if a threshold has been established in a laboratory setting, Denison asked, does it still make sense to talk about such a threshold in a variable human population, "especially in light of cumulative effects and the fact that we are being exposed to multiple chemicals and other types of stressors?"

Another change that occurred over the past several decades is the appearance of a variety of new technologies that can be used to assess chemical hazards and exposures. For example, he said, the emerging high-throughput testing in both in vitro and in vivo settings has the potential to

- address the huge backlog of untested chemicals,
- increase human relevance,
- identify biomarkers of exposure to specific chemicals,
- consider multiple cell types and life stages,
- test at many different doses,
- assess mixtures, and
- inform green chemistry.

There will also be many challenges to putting such new technologies to work, Denison said. How, for example, do we move from an in vitro set of assays to understanding the full implication in a whole organism? Can all potential effects pathways ever be captured? And how do we account for the way exposures take place in the real world, with multiple exposures at different times and chronic exposures?

THE PUBLIC HEALTH APPROACH TO INDUSTRIAL CHEMICAL ASSESSMENTS

Given the presence of so many chemicals that humans may be exposed to, a natural question is how best to ensure human health in the face of these chemicals. William E. Halperin, Chair of the Department of Preventive Medicine at the New Jersey Medical School, described how the field of public health approaches industrial chemicals and industrial chemical assessment.

There is actually no single "public health approach," Halperin said. Instead, public health resembles the proverbial elephant examined by a group of blind men—it can seem to be a rope (the tail), a wall (the side), a pillar (the leg), a tree branch (the trunk), a fan (the ear), and so on, depending on exactly which part is being examined. So, Halperin said, he would illustrate the public health approach to chemicals by describing five different paradigms that could be used in dealing with such chemicals:

the industrial hygiene approach, prevention, surveillance, embeddedness, and dose response.

The first paradigm—the traditional public health approach—is one of anticipation, recognition, evaluation, intervention, and effectiveness. "This is what is taught in industrial hygiene," Halperin said. Anticipation is easy to understand, he said: If you combine a micro-car, a teenage driver, and high speed on a highway, you have to anticipate the problem that you are going to run into.

Sometimes it is not possible to anticipate an issue; in those cases one must recognize it when it appears. Halperin mentioned the case of Ramazzini, who was quoted in about 1710 saying something to the effect of, "I have never visited a nunnery that has escaped the scourges of breast cancer. There must be a connection between the breast and the uterus which escapes the detection of the prossectors [i.e., dissectors]." Ramazzini, who is considered the father of occupational medicine, invented occupational epidemiology in that observation that delayed childbirth put a woman at greater risk of certain adverse effects. Although it was not possible at the time to explain Ramazzini's observation, still it was a classic example of recognition.

Evaluation is somewhat different from recognition. To illustrate, Halperin pointed out that there have been many toxicological studies showing that TCDD (2,3,7,8-tetrachlorodibenzodioxin) is associated with adverse effects. That was recognition. "The evaluation came in the 1980s with NIOSH [the National Institute for Occupational Safety and Health] doing a cohort mortality study of about 6,000 workers highly exposed to dioxin in various occupational situations," which made it possible to then identify the adverse effects associated with TCDD exposure.

Intervention is a product of education, engineering, and regulation, he said, and effectiveness refers to the process of observing whether the steps taken during the intervention were actually effective. "This is the industrial hygiene approach. It is a good broad approach that helps us work on issues of recognition and of hazards."

To illustrate the other paradigms he would be talking about, Halperin began with an anecdote concerning ortho-toluidine, an organic compound used in the production of dyes. Aniline-based dyes such as benzidine have long been known to be carcinogenic. In the 1970s, Halperin said, bioassays suggested that toluidine was actually much more carcinogenic than aniline and that toluidine was associated with bladder cancer in animals. Furthermore, there were a few studies in the 1970s that demonstrated that aniline was not associated with bladder cancer, but that mixed exposures of the aniline-based dyes were.

There were two different groups exposed to toluidine. The first consisted of workers exposed to the chemical in the workplace; the National Occupational Survey in the early 1980s estimated that about 30,000 workers received this occupational exposure. The second group consists of those people exposed to the toluidine found in cigarette smoke, which amounted to tens of millions of people whose exposure was at a much lower level than the workers. Given these occupational and environmental exposures to something that was a known carcinogen in animal studies, there was a reasonable expectation that there would be a resulting health effect.

"That brings us to 1988 when Steve Markowitz, who was an occupational physician working with the Oil, Chemical, and Atomic Workers Union, visited a plant in upstate New York to give a general talk on occupational health," Halperin said. "Workers approached him and asked, Is it meaningful that seven of our cohort have bladder cancer?" The result was that NIOSH was brought in to perform a hazard evaluation. "I was there," Halperin said. "I was on the walk-through." The NIOSH team found signs warning of a suspected carcinogen. "There were literally signs saying there was a suspected carcinogen and that there may well be a problem associated with this plant." The team found out that the suspected carcinogen was toluidine.

After performing an incidence study, the team found that people who had been exposed for 10 years had a risk of developing bladder cancer that was 30 times greater than the general population. There really was a problem.

The effect of all of this work was that ortho-toluidine is now accepted as carcinogenic. And this, Halperin said, is in reality the way that public health has traditionally approached chemicals. The approach is reactive. "We react to observations by astute observers, whether it is Ramazzini or a group of workers at a chemical plant in northern New York State, that there is a problem. We react rather than systematically investigate."

With that example as a touchstone, Halperin described the four other paradigms. The second is the preventive medicine approach, which can be thought of as consisting of primary, secondary, and tertiary prevention. Primary prevention refers to action taken before there is an exposure, he explained. "It is premarket testing so that something doesn't ever get to the workplace. It is substitution or elimination. It is environmental controls. It is personal protective devices. It is all of those kinds of things."

Secondary prevention refers solely to routine periodic screening with the goal of detecting a disease early while it is easier to treat. And tertiary prevention refers to the range of medical care and health care used to respond to a disease once it appears, from drug treatments and surgery to rehabilitation and accommodation. Tertiary prevention has grown immensely in importance over the past 10 years, Halperin said, largely because of the observations of the Institute of Medicine about the number of people who die because of lack of medical care (IOM, 2002, 2003).

In the example of ortho-toluidine, primary prevention would include such things as premarket testing, substitution or elimination, environmental monitoring, environmental controls, and biologic monitoring. "All of this falls in the area, if you will, of toxicologists talking to industrial hygienists," he said. An example of secondary prevention related to toluidine would be screening cytoscopy, and tertiary prevention could include surgery, compensation, and accommodation.

The series of actions in preventive medicine can be thought of in terms of a cascade (see Figure 2-3). "This is how we operate in the industrial setting to reduce adverse effects," Halperin said. "We try our best at these [primary] levels up here. Often times we only find out that there is a problem down here [in the tertiary levels] when there is clinical care made available to seven workers who have bladder cancer."

The third paradigm, surveillance, was fathered by Alex Langmuir at the Centers for Disease Control in the 1950s, although historically it goes back to the 18th and 19th centuries, Halperin said. It is "the systematic, ongoing collection of relevant health-related data—disease, injury, hazard, intervention, etc.—and their constant evaluation and dissemination to all who need to know for the purpose of prevention."

A crucial related concept is the idea of a "sentinel health event." This is a paradigm that was developed at NIOSH in the 1980s, he said. A sentinel health event is "an unnecessary disease or injury, disability, or untimely death which is known to be preventable and whose occurrence serves as a warning signal that preventive or medical care may need to be improved." A closely related hazard, such as a high rate of lead exposure or a low rate of immunization, may also serve in place of a disease or an injury.

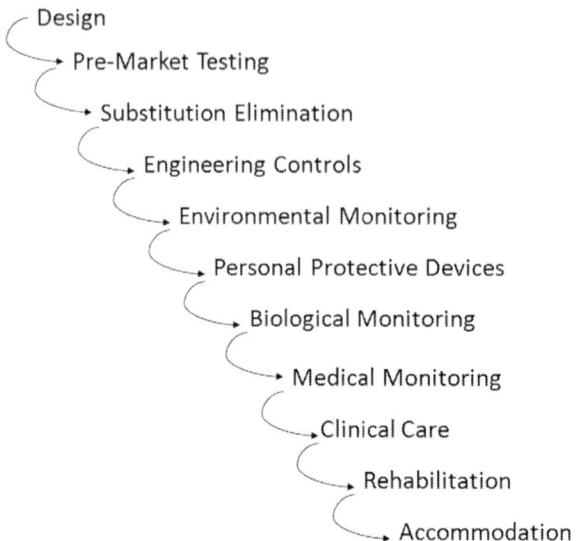

Design
Pre-Market Testing
Substitution Elimination
Engineering Controls
Environmental Monitoring
Personal Protective Devices
Biological Monitoring
Medical Monitoring
Clinical Care
Rehabilitation
Accommodation

FIGURE 2-3 Cascade of prevention: Hierarchy of controls.
SOURCE: Halperin, 2013.

Surveillance serves, in essence, as a feedback mechanism in the cascade of prevention (see Figure 2-4). It is not done to reduce exposure or to ameliorate symptoms directly. Instead, it serves to provide information concerning what is going on and thus to inform various preventive steps, Halperin explained. "Okay, there are seven cases of bladder cancer—what does that mean when we go back to the design of the operation? What does it mean to how we ought to take care of environmental monitoring? Should we be protecting workers better from the exposure?"

The fourth paradigm, "embeddedness," refers to the fact that all of the various actions taken to detect and respond to health issues take place in a larger context. "It is all imbedded in an economics, social, political, regulatory, and ethical matrix," Halperin said. "We all have to work in this matrix."

The fifth paradigm is responding to risk through preventive medicine, which Halperin said could be traced to a book by Geoffrey Rose, *The Strategy of Preventive Medicine* (Rose, 1992). Rose calls for reducing risk throughout an entire population rather than focusing on the small percentage of people within a population who are most affected by a particular risk. Halperin illustrated the idea with a pair of diagrams that could refer to a large number of situations (see Figure 2-5).

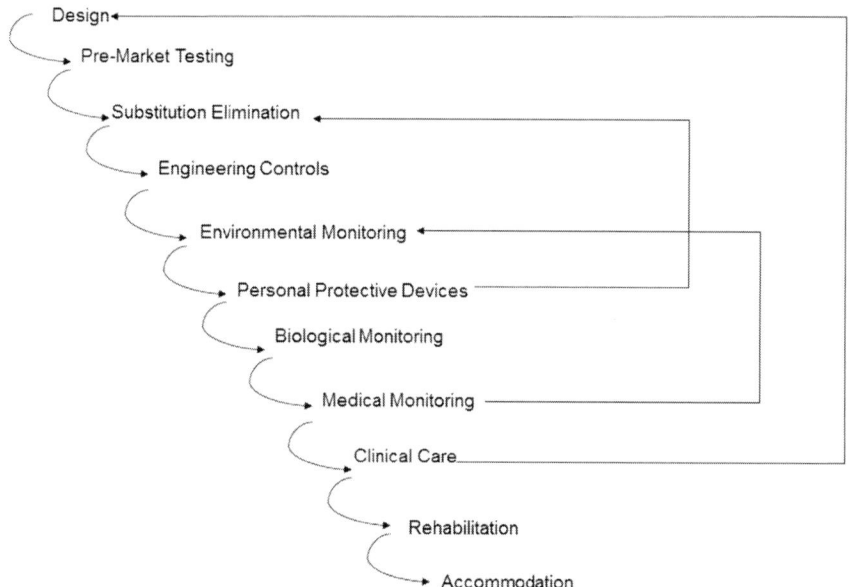

FIGURE 2-4 Cascade of prevention with feedback from surveillance added.
SOURCE: Halperin, 2013.

The graphs could represent a number of things, he explained. They could represent the distribution of systolic blood pressure in the population, for example, or the population exposure to toluidine. In each case there is a group (represented by the hatched region to the far right in Figure 2-5a) that is most at risk and that gets most of the attention in the traditional clinical approaches. "We set a bright line, and we call everybody to the right of it a hypertensive. We see them every two months, and these other people [to the left of the line] we reassure and tell them to go home. The same is true if you think about toluidine."

What Rose proposed was not only to take care of the people at the highest exposure, but to shift the population to the left, as illustrated in Figure 2-5b. The idea is to reduce exposure for everybody, which will have a greater effect than simply working with those people at the highest level of exposure.

This approach has been shown to work with hypertension, Halperin said. Although those people with the highest blood pressures are most at risk of heart attacks, there is an increased risk of heart attack in those with somewhat elevated blood pressures, and because those with moderate levels of hypertension make up a much larger percentage of the population, they account for a large percentage of the overall population risk for heart attack. Those with blood pressures over, say, 180 actually account for a relatively small percentage of the total population effect, Halperin said. Something similar is true for chemical risks, such as with toluidine. In this case there were some 30,000 workers with very high occupational exposure, but there were tens of millions of people with a much lower—but still hazardous—exposure due to smoking. By reducing exposure for everyone—shifting the curve to the left—the preventive strategy seeks to have a greater overall reduction in population risk than would be achieved by focusing on those most at risk.

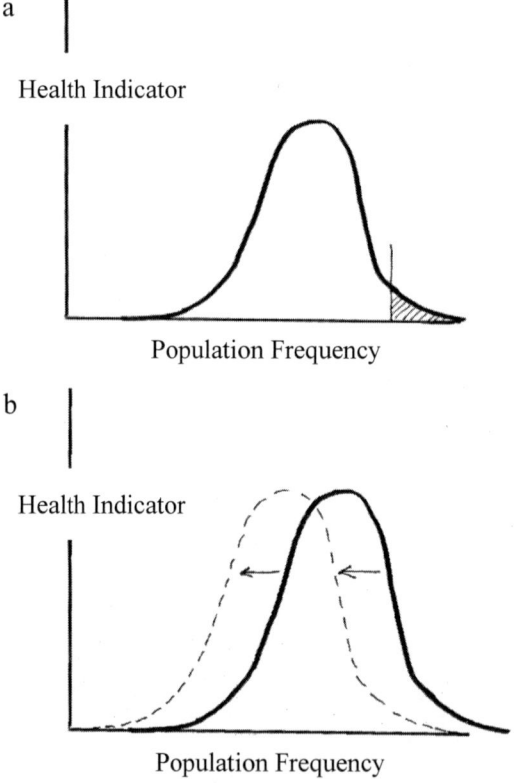

FIGURE 2-5 Responding to risk: (a) traditional approach and (b) preventive medicine.
SOURCE: Adapted from Halperin, 2013.

NATIONAL CONVERSATION ON PUBLIC HEALTH
AND CHEMICAL EXPOSURES

In June 2009 the Centers for Disease Control and Prevention (CDC) along with the Agency for Toxic Substances and Disease Registry (ATSDR) launched the National Conversation on Public Health and Chemical Exposures. The project, which was designed to gather input from a diverse collection of public and private interests, had as its goal the development of ways to ensure that chemicals are used and managed in ways that are safe and healthy for all people. Nsedu Obot Witherspoon, Executive Director of the Children's Environmental Health Network, served as co-chair of the project's leadership council, and she described the National Conversation effort and the Action Agenda it produced to the workshop audience.

The underlying rationale behind the project, Witherspoon said, was that the landscape regarding chemicals in the environment had changed considerably over the previous few years. There had been a growing recognition that there are many different exposure pathways, for example, as well as a recognition that chemicals can lead to a broad range of health outcomes. Biomonitoring had become feasible, and new approaches to toxicity testing had appeared. There had also been movement in the areas of environmental justice, green chemistry, and social media and communications. Thus, it was important to discuss how to effectively proceed in this changed landscape.

As noted, the vision behind the project was to see that chemicals are used and managed in ways that are safe and healthy for all people. Accomplishing that vision would require a number of things, Witherspoon said. It would require accurate information and improved scientific understanding, better policies and practices, and improved prevention, preparedness, and response. It would demand the elimination of inequities and increased engagement by the public and by health care providers. And it would require the development of networks for collaboration and coordination. "We saw this as an opportunity to leverage the work already being done," she said. "This was not at all intended to be a new aspect of work to reduce or eliminate chemical exposures, but rather to leverage and increase partnerships and opportunities, especially during these hard economic times."

A wide variety of partners were involved in the effort. CDC and ATSDR supported the National Conversation initially, and then they worked with a number of organizations to help manage the process.

Resolve, which is an independent, nonprofit facilitator, convened and facilitated the National Conversation leadership council and a number of work groups. "To ensure strong engagement and input from environmental and public health stakeholders and the public, CDC partnered with the American Public Health Association, the National Association of County and City Health Officials, the National Association of State and Territorial Health Officials, and the National Environmental Health Association," Witherspoon said. "Also, Westast's Web dialogue group provided interactive online discussions at key points throughout the project."

The National Conversation had a leadership council and six work groups—on policies and practices, monitoring, serving communities, scientific understanding, education and communication, and chemical emergencies. Witherspoon explained that, during the 2-year process, thousands of members of the public participated through online and in-person forums. The work groups released detailed reports in October 2010. The 40-member leadership council issued the final Action Agenda in July 2011 (National Conversation on Public Health and Chemical Exposures, 2011).[2]

There was a major effort to make sure that the National Conversation included a broad set of perspectives. The leadership council and work groups included representatives from 13 federal agencies, with state, local, and tribal government agencies; national nongovernmental organizations; community and environmental justice groups; academia; industry; and affected communities. "Members of the public participated in over 52 community conversations held across the country and two Web-based dialogues," Witherspoon said. "The leadership council drew heavily from the work group reports and public input in developing the Action Agenda."

The project's main product was an Action Agenda that was intended to offer clear, achievable recommendations that could help government agencies and other actors and organizations strengthen their efforts to protect the public health from harmful chemical exposures. The Action Agenda was divided into seven chapters, each focused on a single priority topic: prevention, monitoring, science, communities, public engagement, health professionals, and emergencies. Each chapter describes the relevant

---

[2] Further information about the National Conversation on Public Health and Chemical Exposures and the final report, *Addressing Public Health and Chemical Exposures: An Action Agenda*, are available at http://www.national conversation.us (accessed March 31, 2014).

public health problem, the challenges to resolving that problem, and the opportunities for new directions. It then offers several featured recommend-ations, along with additional recommendations related to that topic. In total, the Action Agenda offers 19 featured recommendations and 29 additional recommendations. For the remainder of her presentation, Witherspoon focused on the key issues and recommend-ations found in two chapters, Chapter 1 on prevention and Chapter 3 on science.

Chapter 1 is titled "Protecting Public Health by Preventing Harmful Chemical Exposures." It notes that although public health has traditionally emphasized primary prevention—the elimination or reduction of the causes of health problems—this has not typically been the approach taken in the United States regarding exposure to chemicals, and it concludes that, by not focusing on primary prevention, the country is missing opportunities to protect the public from harmful chemical exposures, said Witherspoon.

To improve the situation, the chapter offers several featured recommendations. She noted that it calls on government agencies to prioritize the reduction of exposure to harmful chemicals. All levels of government are encouraged to provide policy incentives, to invest in research and development, to develop enhanced hazard screening methods, and to disseminate information for personal decision making.

The chapter also calls on Congress to reform TSCA and encourages states to enact similar legislation. And indeed, Witherspoon said, such actions have been taking place at the state level for many years now. California, Maine, Massachusetts, Minnesota, and Washington have all implemented toxics use reduction legislation and initiatives. At the federal level there have been various efforts to reform TSCA, although none has yet come to pass.

Another featured recommendation in the chapter is to make improved protection of children's health a priority. Witherspoon noted that government agencies are asked to require explicit consideration of children's unique vulnerabilities, susceptibilities, exposures, and develop-mental stages as well as of the places where children, live, learn, and play.

Chapter 1 also has several additional recommendations, explained Witherspoon. They call for increasing the emphasis on public health principles and precaution; developing standard scientific criteria and protocols for applying a precautionary approach to chemicals; strengthening protections of workers' health; ensuring that industrial and federal facilities comply with environmental health regulations, laws, and policies; and developing an overarching paradigm for assessing risk.

Chapter 3, which is titled "Achieving a More Complete Scientific Understanding of Chemicals and Their Health Effects," concerns the importance of knowledge and understanding concerning chemical toxicity, modes of action, sources of exposures, and potential adverse health effects. The United States has undertaken significant research efforts, Witherspoon said, yet we lack critical information on the health effects of chemicals, including low-dose, multiple, and cumulative exposures; on individual susceptibility and intolerance including, but not limited to the interplay between genes and environment; on community vulnerability and disproportionate effects from past exposures; and on the effectiveness of interventions to protect public health.

The featured recommendations in Chapter 3 are aimed at enhancing scientific methods, tools, and knowledge. They encourage the expanded use, further development, and validation of modern molecular biology techniques, computational systems biology, and other novel approaches, said Witherspoon. Government agencies are encouraged to identify the data needed to fill gaps in the scientific understanding of health risks of chemicals and also to prioritize chemicals of concern for further assessment of exposures and safer alternatives.

The enhancements that the EPA made to its Integrated Risk Information System (IRIS) in summer 2013 are just one example of recent actions consistent with Chapter 3's recommendations, Witherspoon said. "The recent IRIS enhancements are intended to improve the scientific foundation of assessments, increase transparency in the program and the process, and allow the agency to produce more IRIS assessments each year. Standard protocols and tools to characterize human exposures across the life cycle of chemical products and across human life stages are also prioritized."

Chapter 3 has five additional recommendations as well, explained Witherspoon. They address coordinating and improving accessibility of databases, understanding individual susceptibility to chemical exposures, defining gene–environment interactions related to chemical and environmental exposures and social and lifestyle factors, identifying the health impacts of indoor air quality during fetal and child development, and evaluating ATSDR's scientific methods.

In the 2.5 years between the release of the report and the workshop, there were a number of examples of agencies acting in ways that were consistent with the report's recommendations, Witherspoon said. For example, the National Center for Environmental Health at CDC convened a workshop to examine ATSDR's scientific approaches, which reflected Recommendation 3.8. And in the chapter on communities, it was recommended that the ATSDR be more focused on community concerns. "Since then," Witherspoon said, "they have been creating geographic branches to ensure that all work is happening in a particular area and is more coordinated and also moving staff from their ATSDR headquarters to more of the local community levels and hiring more staff in the regions."

In June 2012, she said, the National Conversation Network was formed with the intention of encouraging further progress toward implementing the report's recommendations and identifying opportunities for collaboration. Furthermore, the Environment Section of the American Public Health Association has a work group specifically organized around the Action Agenda. In particular, she added, there seems to be a great deal of interest right now in including environmental considerations into undergraduate, graduate, and health professional curriculum development and training. In addition, the call for TSCA reform is "resonating very loud and clear."

In conclusion, Witherspoon said that the National Conversation's Action Agenda offers an effective roadmap for leveraging current partnerships and dealing with the lack of resources available. "This is a great time for us to take a step back to take an assessment of some aspects and initiatives that haven't been so effective and think about re-routing ourselves," she said. "This is a roadmap that I encourage all of us to spend some time reviewing and acting on."

REFERENCES

ACC (American Chemistry Council). 2014. *U.S. chemical production expanded in December: 2013 ends on a high note.* Available at http://www.american chemistry.com/Media/PressReleasesTranscripts/ACC-news-releases/US-Chemical-Production-Expanded-in-December-2013-Ends-on-a-High-Note.html (accessed March 30, 2014).

EPA (U.S. Environmental Protection Agency). 2014. *TSCA chemical substance inventory: Basic information.* Available at http://www.epa.gov/oppt/existing chemicals/pubs/tscainventory/basic.html (accessed March 30, 2014).

GAO (U.S. Government Accountability Office). 2013. *Toxic substances: EPA has increased efforts to assess and control chemicals but could strengthen its approach.* Washington, DC: GAO.

Goldman, L. R. 2002. Chapter 17: Toxic chemicals and pesticides. In *Stumbling toward sustainability*, edited by J. C. Dernbach. Washington, DC: Environmental Law Institute.

Halperin, W. E. 2013. Public Health Approach to Industrial Chemical Assessments: 5 Paradigms. Presentation at the Institute of Medicine Workshop on the Identifying and Reducing Environmental Health Risks of Chemicals in Our Society, Washington, DC.

IOM (Institute of Medicine). 2002. *Care without coverage: Too little, too late.* Washington, DC: National Academy Press.

IOM. 2003. *Hidden cost, value lost: Uninsurance in America.* Washington, DC: The National Academies Press.

National Conversation on Public Health and Chemical Exposures. 2011. *Addressing public health and chemical exposures: An action agenda.* Atlanta, GA: Centers for Disease Control and Prevention, Agency for Toxic Substances and Disease Registry.

NRC (National Research Council). 1983. *Risk assessment in the federal government: Managing the process.* Washington, DC: National Academy Press.

NRC. 2009. *Science and decisions: Advancing risk assessment.* Washington, DC: The National Academies Press.

Paracelsus. 1531. Preface to the Coelum Philosophorum.

Rose, G. 1992. *The strategy of preventive medicine.* Oxford: Oxford University Press.

Safer Chemicals Healthy Families. 2012. *Chemicals and our health: Why recent science is a call to action.* Available at http://saferchemicals.org/PDF/chemicals-and-our-health-july-2012.pdf (accessed March 30, 2014).

SOCMA (Society of Chemical Manufacturers and Affiliates). 2014. *Myth versus fact about chemicals in commerce.* Available at http://www.socma.com/GovernmentRelations/index.cfm?subSec=26&articleID=3259 (accessed March 30, 2014).

UNEP (United Nations Environment Programme). 2013. *Global chemicals outlook: Towards sound management of chemicals.* Geneva: UNEP.

Ward, E., C. DeSantis, A. Robbins, B. Kohler, and A. Jemal. 2014. Childhood and adolescent cancer statistics, 2014. *CA Cancer Journal for Clinicians* 64(2):83–103.

Wilson, M. P., M. R. Schwarzman, T. F. Malloy, E. W. Fanning, and P. J. Sinsheimer. 2008. *Green chemistry: Cornerstone to a sustainable California.* Berkley, CA: Centers for Occupational and Environmental Health, University of California.

3

Current Regulatory Approaches to Dealing with Industrial Chemicals

How should the potential hazards posed by chemicals, both environmental hazards and health risks, be assessed and dealt with? Several speakers at the workshop described current regulatory approaches to dealing with industrial chemicals and discussed both their successes and their shortcomings.

THE TOXIC SUBSTANCES CONTROL ACT

Lynn R. Goldman devoted much of her workshop presentation to a discussion of the Toxic Substances Control Act (TSCA) of 1976,[1] the key federal law governing hazardous chemicals in the United States. "This was the first bill that was ever written anywhere in the world to attempt to regulate toxic chemicals," she noted. The act excludes food and food additives, drugs, cosmetics, pesticides, tobacco and tobacco products, and a few other categories of chemicals from its coverage, but otherwise it applies to all chemicals used in the industrial context and the commercial context as well as in most consumer products.

Two Universes

Wendy Cleland-Hamnett, Director of the U.S. Environmental Protection Agency's (EPA's) Office of Pollution Prevention and Toxics, offered some details about the act and the challenges of implementing it. TSCA divides chemicals into two "universes," she explained—those chemicals that existed at the time the statute was passed and implemented

---

[1] Toxic Substances Control Act of 1976, Public Law 94-469, 94th Congress.

and those new chemicals that came onto the market after that. After the passage of the act, one of the first things that the EPA did was to collect information to compile an inventory of chemicals that were in commerce at the time, that is, in the late 1970s. The inventoried chemicals numbered about 62,000. "After that inventory was completed, Cleland-Hamnett said, "any chemical that was subject to TSCA jurisdiction that was not on that inventory of existing chemicals had to come through our new chemicals program."

In particular, TSCA requires a premanufacture review of any new chemicals. "A company that wants to market a chemical has to notify us that they intend to do that at least 90 days before the chemical goes on the market," she said. The company is required to provide certain information about the chemical, but it is not required to generate toxicity data in order to submit a premanufacture notice to the agency. The company is required to give the EPA any information it has in its possession at the time of the notification, but not to generate any information, she explained.

Over time the EPA has developed an approach to evaluating new chemicals that relies on quantitative structure activity analysis and modeling, both on the toxicity side and on the exposure side, explained Cleland-Hamnett. After receiving notification from a company of its intention to begin manufacturing a new chemical, the EPA has 90 days to evaluate the chemical and develop next steps. Once the 90 days expire with no EPA action, the company can go ahead and begin its production.

In order to take action on a new chemical about which it has some concern, the burden is on the EPA to show either that the chemical may present an unreasonable risk or that the chemical may be produced in substantial volumes and there may be substantial exposure, Cleland-Hamnett said. "We have to make a case for toxicity or exposure." If the agency makes that finding, it has the authority to require testing of the new chemical substance, and risk-reduction provisions can be put into place while that testing is being developed. "We can do that by administrative order," she said, "but most of the actions that we have taken since the beginning of the program have actually been via consent order that we negotiate with the companies that are submitting the notifications."

Sometime in the 1980s after the program had been in place for a while, the EPA also started to routinely issue significant new use rules for those chemicals where it had negotiated consent orders. The consent order applies only to the company providing the notification, Cleland-

Hamnett explained, and once the chemical goes on the inventory, even though it is subject to restrictions for the original company, those restrictions would not apply to other companies. Thus, the agency puts a rule in place to apply similar restrictions to any other company that wants to proceed with manufacturing that chemical.

Since the new chemicals program began, there have been more than 22,000 new chemicals that came through the new chemicals review process and were added to the inventory (GAO, 2013). The EPA has actually reviewed about twice that many premanufacture notifications or other new chemical notifications, Cleland-Hamnett said, but only about half of the notifications it reviews actually end up being commercialized. The companies decide not to pursue commercialization for various reasons—"sometimes, I think, because we raise issues about them," she said. And that is the new chemicals side of the TSCA universe.

For the existing chemicals side, which consists of the 62,000 chemicals that were identified back in the late 1970s, there is no mandatory review process under TSCA, she said, although the act does give the EPA authority to collect data on such things as production volume and use (GAO, 2013). "We do now have in place a chemical data reporting rule where we periodically every 4 years collect information on a subset of the chemicals that are part of the existing chemicals universe."

The EPA does have some authority under TSCA to require, through a rule-making process, the testing of existing chemicals, but there are certain findings that the EPA needs to make in order to require that testing. As Goldman noted in her presentation, the factors the EPA most often uses to decide to require a chemical to be tested include the quantities manufactured and released, the numbers of people exposed occupationally and nonoccupationally, similarity to a substance known to pose unreasonable risk, the existence of data concerning environmental or health effects, the quantity of information to be gained, and the availability of testing facilities and personnel.

"Over the 30-plus years of TSCA, I think we have required testing on over 200 chemicals," Cleland-Hamnett said. "We go through notice and comment. We go through the executive order reviews. It is a fairly lengthy process." It takes an average of about 5 years to get the final rule in place, and then the necessary data are collected, which can take several more years depending on the types of testing required, she said. "Then you have the review of the data. It is not what you would call a really responsive process."

The EPA negotiated a voluntary chemical hazards screening program with industry in 1998 titled the High Production Volume Challenge Program,[2] Goldman noted. According to the agreement, chemicals with a production of more than 1 million pounds a year are screened using the Organisation for Economic Co-operation and Development Screening Information Data Sets program,[3] and the screening information is provided to the EPA and to the general public. Approximately 2,200 chemicals have been sponsored by companies that participate in that voluntary program (EPA, 2012).

Under TSCA, Goldman said, the EPA has the responsibility for managing risks of certain chemicals. "Congress actually told EPA how to manage polychlorinated biphenyls, asbestos in certain uses, radon, lead, mercury export, and formaldehyde in wood products," she said. But other chemicals must be assessed using risk/benefit criteria and analytic processes to identify the "least burdensome" approach to managing the risks. Additionally, TSCA provided the EPA with the ability to refer certain identified chemical risks to other regulatory agencies. In the 1980s, the EPA referred some chemicals to the Occupational Safety and Hazard Administration and it referred dioxin in food packaging to the Food and Drug Administration for management by those agencies. "EPA itself took action to manage certain risks for chlorofluorocarbons, dioxin, asbestos, and hexavalant chromium," she said, "but their asbestos rule was overturned by one of the circuit courts [see Box 3-1]. Since then, EPA has done very little." In May 2010 the agency submitted a list of proposed "chemicals of concern" to the Office of Management and Budget (OMB) (EPA, 2009), but the agency ran into obstacles at the OMB and withdrew the list in September 2013.

By contrast with the situation for existing chemicals, many risks from new chemicals have been addressed by the EPA since the passage of TSCA. Between 1976 and 2012, a total of 2,180 new chemicals were subjected to significant new use rules, or SNURs (GAO, 2013). The rate of use of these rules has been growing; about 25 percent of the SNURs were issued between 2009 and 2012 (GAO, 2013). Additionally, Goldman said, "as a guesstimate [there are] maybe around 5,000 chemicals that EPA has one way or another managed, including by not approving

---

[2] Further information on the High Production Volume Challenge Program is available at http://www.epa.gov/hpv (accessed March 30, 2014).
[3] Further information on the Screening Information Data Sets program is available at http://www.epa.gov/opptintr/sids/pubs/overview.htm (accessed March 30, 2014).

them." She pointed out that industry will often withdraw a new chemical from review to avoid a rejection.

The management of chemical hazards is a tough area for the EPA, Goldman said. Part of the reason is the circuit court decision on the asbestos ban, which has given the EPA a very high bar to surmount in order to manage risks. But another part of the reason is the difficulty of understanding exactly what the public health hazards are that are associated with chemical risks generally.

Companies have various reporting requirements under TSCA. For example, if testing shows significant risk for an existing chemical or new

---

BOX 3-1
The Asbestos Court Decision and Its Legacy

In the late 1980s and early 1990s, the EPA attempted to use its regulatory authority for existing chemicals under the Toxic Substances Control Act to try to ban a number of uses of asbestos. "The agency spent something like 10 years and many, many millions of dollars compiling the case to do that," said Wendy Cleland-Hamnett, Director of the EPA's Office of Pollution Prevention and Toxics. However, once the final rule was issued, it was challenged in court, and eventually the Fifth Circuit Court overturned most of the rule. "The court found that despite the fact that we were talking about asbestos and despite the fact that there was a very large record compiled for that case, we still had not met the cost–benefit requirements for banning products containing asbestos," Cleland-Hamnett explained. The agency had not met requirements of demonstrating that it had looked at every potential use of asbestos and that for every potential use it had identified all of the alternatives and what the implications of switching to those alternatives would be. In particular, she said, "the agency had not made the case that we were taking the least burdensome approach to dealing with those products containing asbestos."

The most important outcome of that court decision was it set a very high bar for the EPA to manage risks, said Lynn Goldman, Dean of the George Washington University School of Public Health and Former Assistant Administrator for Toxic Substances at the EPA. Two things in particular about that ruling have made it difficult for the EPA to regulate hazardous materials, she said. First, the court ruled that the EPA needed to do a much more extensive substitution analysis to prove it had really selected the best approach to managing the risks. Second, the court showed a very strong preference to "end of pipe solutions," that is, for proving that the imposed regulation represents the least burdensome approach, which can be very difficult to prove. Since that ruling, the EPA has not used its regulatory authority to restrict the use of any existing chemical.

chemical, it must be reported. Even a research-and-development chemical that shows such risk must be reported, Goldman said.

Because companies must provide information to the EPA, there are provisions in TSCA for the protection of confidential business information. "Those are important provisions in protecting proprietary interest in chemicals," Cleland-Hamnett said. "Particularly in the area of new chemicals where innovation is going on, there is a societal interest in protecting the value of that information." On the other hand, she added, because the EPA has not always made sure that companies really needed confidentiality concerning particular information, particularly years or even decades after the confidentiality had initially been granted, "there were a lot of things that were being kept confidential that perhaps really didn't merit that protection." In response to that situation, Cleland-Hamnett said, about 5 years ago the EPA started a program to identify unwarranted claims of confidentiality so more of the information that had been collected under TSCA could be put into the public domain.

How the EPA Is Implementing TSCA Today

Cleland-Hamnett spent the final part of her presentation describing what the EPA is doing now in its implementation of TSCA. In particular, she focused on three interrelated aspects of the program: risk assessment and management, access to data, and safer chemicals.

In the first area, she said, "we are trying to strengthen and revitalize and basically build a program to assess and manage existing chemicals." In the past the agency has never been able to look at more than 5 to 10 chemicals per year. "Clearly, that sort of pace is not going to get us where we want to be—or where I think the country wants us to be—on chemicals management."

In prioritizing chemicals for risk assessment, the agency looks for six characteristics: potentially being of concern to children's health, either through reproductive or developmental effects; having neurotoxic effects; being persistent, bioaccumulative, and toxic; being a probable or known carcinogen; being used in products to which children might be exposed; and having been detected in biomonitoring. Using these characteristics, the EPA identified a Work Plan of 83 chemicals for review and risk assessment. In January 2013 the agency released draft risk assessments for 5 chemicals for public comment, to be followed by peer review, and 2 months later it announced 23 chemicals that would be assessed in 2013, including 20 flame-retardant chemicals. "We issued four draft risk assessments earlier this year," Cleland-Hamnett said, "and we will be issuing additional ones. We are going through peer review on those draft risk assessments right now." If an assessment indicates potential concerns, the EPA will evaluate and pursue appropriate risk management efforts; if it shows negligible risks, the agency will conclude its work on the chemical. The EPA will continue to conduct risk assessments on the remaining Work Plan chemicals and add additional chemicals to the Work Plan if warranted.

In the area of access to data, she said, the EPA has over the past several years taken a range of significant steps to increase the public's access to information and reduce confidential business information (CBI) claims, and it will continue efforts to improve the accessibility and usability of chemical data. "Also, in September we released a new chemical information portal that we will be building over time, but we are very excited about it," she said. "The website is epa.gov/chemview. We have data on over 8,600 chemicals in there now . . . and we will be working over the coming years to add information. Our goal is to basically have everything in there by 2017."

In the area of promoting safer chemicals, the EPA has two programs: a green chemistry program that promotes the development of safer chemicals, and Design for the Environment, a program that recognizes products that contain safer chemicals. There is a "very robust set of criteria" used to evaluate chemicals for the Design for the Environment label, Cleland-Hamnett said, and companies that want to get that label for some of their products must develop a dossier and submit it to a third party for review. If the substance passes that review, it comes to the agency for recognition. So far the EPA has recognized nearly 3,000 products with the Design for the Environment designation. "About a year ago," she added, "we also took the chemicals that were in those recognized products and separately listed them on a Safer Chemical Ingredients List by functional class. If someone is looking for a chemical that meets a functional need that we have recognized as a safer chemical, they can look to that list." The agency is also working to encourage companies to submit chemicals to be included on that safer chemical ingredient list.

Despite the obstacles that the EPA faces, the job of assessing and managing chemicals is an important one, Goldman said. It has been estimated that chemical risks lead to at least $76.6 billion per year in medical care costs, not including the costs of occupational diseases. "I don't know if this is actually a credible number," she said. "I think that it is likely to be an underestimate, given the fact that we have such incomplete knowledge of chemicals in commerce."

"I think this is a very important area," Goldman concluded. "It is a challenging area not just for the EPA, but really for all of us in public health and environmental health."

Perceived Concerns with TSCA

Over the course of the workshop, several individual speakers talked about what they saw as problems with TSCA—problems that make it difficult to make sure that the U.S. population and environment are protected from harmful chemicals—and what the causes of those problems are as well as possible solutions.

For example, Cleland-Hamnett said that, generally speaking, many of the problems with TSCA derive from the fact that there is no mandatory program to review those existing chemicals. "As anybody who has been in Washington for any time would tell you, if there isn't a mandate to get it done, it is often not going to get done," she said. "You are at the back of the line behind statutory deadlines for policy-level attention and for budget attention. . . . That is one of the things that we have struggled with."

Another key issue, Cleland-Hamnett said, is the existence of legal and procedural hurdles to the limiting or banning of chemicals. One of the biggest is the court-established hurdle from the asbestos case (see Box 3-1). The court ruled, in essence, that the EPA needs to show that it is taking the "least burdensome" approach in regulating a particular chemical. "Up until that asbestos rule," Cleland-Hamnett said, "the agency had used its existing chemical regulatory authority five times for a number of chemicals through the 1980s. After that asbestos situation we haven't used it at all, from the early 1990s until 2013."

Richard Denison spent much of his presentations discussing the weaknesses of TSCA and "why so many of us believe that we need a new law." One of the biggest problems, he said, is that the original act grandfathered in tens of thousands of chemicals. "Essentially TSCA created a presumption that those chemicals were safe," he said. "Until and unless EPA found really compelling evidence of harm, those chemicals were basically deemed to be off limits." And this in turn has created what he termed "an odd disincentive to create new information." Because the default is that if the information does not exist or is uncertain, no action is taken. Hence, companies avoid generating additional information because it could provide the evidence the EPA needs to act on a chemical.

Another problem, he said, is the high hurdle that must be cleared to require testing. He pointed to Cleland-Hamnett's observation that it takes an average of 5 years to go through the rule-making process to require testing as evidence of just how difficult it is for the EPA to order such tests.

Other problems are the high level of proof of harm required before an action can be taken to regulate chemicals and the fact that the burden of proof is on the EPA to show a substance is dangerous rather than on a manufacturer to show it is safe. This is a very marked difference from the way the United States regulates chemicals like pesticides and drugs. Whereas those chemicals are designed to be biologically active, Denison said, we now know a lot of chemicals that were never designed to be biologically active actually are.

The result, he said, is the situation that other speakers had already described: 62,000 chemicals grandfathered in by the original act, only about 300 chemicals required to be tested in the years since the act was passed, only 5 chemicals that have been regulated in limited ways, and 22 years since the last time the EPA tried (and failed) to regulate a chemical: asbestos.

The situation with new chemicals may not be as bad as that with existing chemicals, but it is not particularly good, either, Denison said. "I think it has been popular to argue that the new chemicals program is the success story under TSCA. I beg to differ." Certainly more has been done with new chemicals, he said, in part because of TSCA's mandate that the EPA review chemicals that are coming onto the market, but there are still a number of concerns that need to be addressed.

First, the default position on new chemicals is "no data, no problem." Because the EPA does not have the ability to impose up-front data requirements on new chemicals, he said, the vast majority of chemicals that come into the agency have no health data. One estimate holds that as many as 85 percent of new chemicals that come into the EPA do not have health data. "This is unlike virtually every other developed country in the world."

As a result, the EPA has been forced to rely heavily on limited prediction models and estimates. These have some strengths, Denison said, but they also have a number of limitations. "For example, such models basically do not exist for most mammalian chronic toxicity end points, which is where a lot of the action is today."

In essence, Denison said, the EPA faces a Catch-22: To require testing, the agency must first have data that suggest potential risks or very high exposures, but it often has no way of getting those data without requiring testing.

Furthermore, the EPA typically gets only one bite of the apple, he said. That is, it gets only a single review opportunity, and it must make assumptions about what that chemical might be used for and what its volume of production is likely to be out into an indefinite future. "This is really their one opportunity to look at it," he said, "unless they . . . promulgate a significant new use rule, which they do in only about 6 percent of the cases of new chemical reviews."

Another problem, Denison said, is the "black box nature" of the EPA's new chemical program. "Wendy [Cleland-Hamnett] mentioned the use of consent orders, but trying to find them is almost impossible," he said. "When you do find one, 90 percent of the stuff in it is blacked out, including things like how long the company has to deliver the data that is required in the consent order. It is a real frustration for those of us outside looking in."

A final problem with the current paradigm, he said, is that the burden is on the EPA to find affirmative evidence of harm in order to act to control a new chemical. As an example, he mentioned carbon nanotubes. "EPA is able to require testing now for inhalation toxicity of those because there are studies out there that suggest that is a real concern with the inhalation of these materials," he said. "They can't require any other testing because there is no evidence out there that suggests it. These brand new materials are coming onto the market, and only where EPA already has evidence of harm can it actually require some testing."

In addition to the problems with the current testing paradigm, Denison said, assessing risk is also hobbled by a lack of data about uses and exposure. For example, information on use is required only for chemicals that exceed the threshold of 100,000 pounds per year at a site. In 2012, information on consumer and commercial use was reported for about 3,600 chemicals. In 74 percent of the cases the manufacturers indicated that at least one of the six basic reportable data items was "not known or reasonably ascertainable." The manufacturers do not necessarily always know this use information, he said, but its absence does indicate how poor the system is performing in providing a full picture of the use of these chemicals.

There is also limited availability of hazard data. In recent years the EPA has looked for hazard data on about 9,900 of the highest-volume chemicals, including pesticides and air and water pollutants. Only about one-quarter of them have detailed toxicology information, while more than 40 percent do not have even limited hazard data.

Denison concluded by saying that it is time for a paradigm shift. "In my view we need to be moving to a system that requires affirmative evidence of safety as a condition for chemicals entering the market or staying on the market. That is what the debate around TSCA reform is all about."

Industry Perspective

Michael Walls, Vice President of Regulatory and Technical Affairs at the American Chemistry Council (ACC), offered some thoughts about regulating chemicals from the perspective of the chemical industry.

To begin, he said he thought that, in talking about chemical risk and management, it is important to think about the message that is being sent. For example, the rapidly growing use of chemicals throughout the world is evidence of economic growth. "The World Bank says between 1970 and 2010 the world economy grew three times," he said. "This means we have a more affluent population. They are starting to demand the goods and services that improve the quality of their lives." That is the story behind the growth of chemical production.

To ensure that chemicals are safe, the industry needs to provide information about them, he said, "but I think we want to make sure the information we are putting out there is meaningful, is relevant, and can be understood by those who use it in a particular case." Generally speaking, he said, there is a wide range of points at which information about chemicals is relevant. "One is clearly in the value chain. Business is making decisions about what products they use to confer a particular characteristic to get to a particular product or result. Another is what information is sent to the government—information EPA is relying on to make decisions. Then there is public communications as well."

In risk management, information is needed for various tasks: to understand what the alternatives are to a particular substance, to understand what the costs and benefits are of a particular regulatory path, and for use in compliance and enforcement, said Walls.

There are many issues concerning this information, he said: Who gets the information? When do they get it? What are they getting? Is it use information? Is it ingredient information? Is it process information? What about the possibilities of unfair or wrong conclusions to be drawn from that information? What is the societal impact of those conclusions? How do you provide information in an efficient and effective way?

Walls listed several examples of successes in the area, although he noted that none are "unqualified successes." There have been some successes in voluntary industry initiatives, he said. "In partnership with the Environmental Defense Fund, the chemical industry joined in a program, the High Production Volume Challenge Program, to ensure that screening-level information was available for the roughly 2,000 chemicals that represent the 95 percent of the volume of chemicals in U.S. commerce. That was a general success, I think."

There have also been industry ingredient-disclosure programs. "ACC members and our colleagues in the International Council of Chemical Association have produced and have made available over the Web about 2,200 safety summaries on major chemicals in commerce," he said. "These aren't intended to be highly technical documents. They are intended to be accessible to the public to generally describe some of the hazards, uses, and exposures of chemicals out there."

Another area of success are nongovernmental standards that have been created related to chemical safety. One example is the NSF International/American Chemical Society Green Chemistry Institute®/ American National Standards Institute (NSF/GCI/ANSI) Greener Chemicals and Processes Information Standard,[4] ANSI-355, which was developed by a group of stakeholders to encourage business-to-business discussions about which safety or sustainability considerations are most important and to help customers encourage a dialogue with chemical manufacturers. Walls noted that the lack of information on implementation of ANSI-355 raised questions about the degree of success with this standard to date.

There are a number of government regulatory actions that Walls said are generally considered to be successes. The Occupational Safety and Health Administration hazard communication standard makes some necessary information available to employees, for example. And there are a growing number of state programs that also make information on chemicals more available.

Walls also said he thinks that the EPA's new chemicals program has generally been a success. "EPA doesn't have every bit of information about every new chemical that comes in, but that doesn't mean that EPA

---

[4] Further information on the NSF/GCI/ANSI Greener Chemicals and Processes Information Standard is available at http://www.nsf.org/services/by-industry/ sustainability-environment/green-chemistry/nsf-gci-ansi-355 (accessed March 31, 2014).

is operating in an information vacuum. They have got 37 years of experience. They have got some great scientists over there in the program that are able to make a reasoned judgment about whether or not a new chemical can be put on the market." The key point to keep in mind when talking about information on chemicals, he said, is that there is not just a single source for such information. There are many sources. "We have got to consider ways in which we can best tap those various sources."

There are a number of areas for potential improvement, Walls said. "One is when and what information should be disclosed in the value chain. If safer chemistry is the goal I think we have to be clear about what is safer. I think we have to create mechanisms that encourage that flow of information [through the value chain]." The European Commission has done that to some extent in Europe, he said, and it is possible that TSCA reform could have a similar effect in the United States. "I think we need to have a discussion of what standard we want for public information on chemicals. What needs to be disclosed? Should that information be tailored to specific uses? Are we providing the public the right tools to interpret that information?"

There also needs to be careful consideration given to the basis for government action, Walls said. "We need to have a clear discussion on how the government decides when it needs more information and then how it goes about executing those decisions." In particular, there needs to be up-front justification of claims on confidential business information, he said. "This is a key issue in TSCA reform."

Looking to the future, Walls noted that the current proposed legislation in Congress aimed at reforming TSCA, the Chemical Safety Improvement Act (CSIA) of 2013,[5] would provide the EPA new authority to obtain information on new and existing chemicals, and it would require the EPA to make safety decisions solely on the basis of health and environmental considerations, not on the basis of a cost–benefit analysis. The Act would also mandate up-front substantiation of CBI and public disclosures, which would be a significant change from the current law. It helps draw a better picture of how chemicals are moving in commerce in the United States, he said. "Besides having manufacturers being required to report, processors would also have reporting requirements."

---

[5] Chemical Safety Improvement Act of 2013, S. 1009, 113th Congress, 1st session (May 22, 2013).

Finally, Walls commented on keeping TSCA focused on industrial production of chemicals. "Given the breadth of information sources out there and the purposes to which we can put chemical information, I think the key point is not to vest in TSCA the requirement to provide all information on all chemicals for all purposes," he said. "I think we have to look at TSCA predominantly as a mechanism to regulate the industrial production and use of chemicals." At least in this country, we still need statutory regimes that look to regulate specific use (such as pharmaceutical or pesticides) beyond that, he noted.

THE EUROPEAN COMMISSION APPROACH

Canice Nolan, Senior Coordinator for Global Health for the European Commission Directorate General for Health and Consumers, spoke to the workshop about the European Commission's perspective on chemicals and chemical risk. Much of his presentation was focused on REACH, which refers to the European Commission Regulation 1907/2006 on the Registration, Evaluation, Authorisation, and Restriction of Chemicals.[6]

Nolan began by commenting that he has generally found the interactions between the U.S. and European agencies responsible for regulating chemicals to be helpful for everyone involved. "We have different perspectives, different backgrounds, different legislations and rules and so on that we work under," he said. "In spite of the differences we do in fact have many of the same challenges facing us."

The chemical industry accounts for a large part of the European economy. With 25,000 companies employing 1.7 million people, it is the third-largest manufacturing sector in the European Union (EU), said Nolan. Although the EU's share of the worldwide market for chemicals has been declining, thanks mainly to the increase in China's share, the total sales of the European chemical industry have been growing.

While the manufacture and use of chemicals is crucial to the health of the European economy, there have been increasing concerns about the negative effects of some of those chemicals. The World Health Organization has estimated that 15 percent of all deaths in the EU are due to environmental stressors (EEA, 2010). "I am not saying that they

---

[6] Further information on REACH is available at http://ec.europa.eu/enterprise/sectors/chemicals/reach/index_en.htm (accessed March 31, 2014).

[the stressors] are chemicals," Nolan said, "but chemicals are certainly part of it." In addition, there is great political pressure to decrease the level of pollution and chemicals in the air and water beyond what has been mandated by existing chemical legislation. Thus, it has become important to think in terms of a bigger picture and not look just at safety aspects, he said. "REACH is the key to achieve this."

Discussions about REACH began in 1999 or 2000, Nolan said, in large part because of the recognition that "there may be 80,000-plus chemicals out there, and we just had information on a small percentage of them." An impact assessment that was done in 2002–2003 found that instituting the proposed REACH regulations would be worth about €25 billion ($35 billion) over a 30-year period just in terms of the health benefits and would be worth about the same for the environmental benefits. However, the assessment also found that the benefits would not start appearing until at least 10 years into the program. "It is still the early days," he said. "REACH was adopted in 2006."

By way of comparison, Nolan described the EU pesticides legislation passed in 1991. "There were 1,000 active substances on the market at that stage, and it was 1999 before the first decision on a substance was taken," she said. "It took that long to prepare the ground, lay out the framework, get industry to prepare the dossiers, and evaluate the dossiers." That job is now done, he said. "About half of the active substances that were on the market in 1991 are now no longer on the market, but not necessarily for health and safety reasons." There were many substances that were produced in very low volumes that industry decided were not worth continuing to produce, she explained. It would have cost too much to go through the necessary processes to keep those substances on the market.

This was a crucial difference between the original pesticides legislation in the EU and the later legislation: In the earlier legislation the burden of proof was on the European Commission to prove that a substance was unsafe. "In 25 years we managed to do this for about 12 substances," Nolan said. "The final decisions weren't taken at the commission level. We had to make a proposal to the council. It would be as if Congress would have to take a decision on these substances. It was a heavy, heavy procedure." With the later legislation, the burden of proof was on a company to prove that a chemical it manufactured was safe. "We flipped it around and said if you are not on the positive list by 2001, you are off the market. Suddenly everybody started scrambling."

The approach used by REACH is similar, but it applies to many more substances. "It basically saves tax payer dollars by putting the onus on industry to show safety," Nolan said. "It also enables the authorities because it is not the commission that will control what is happening out in the field, it is the authorities in the member states. They now, through REACH, have legislative powers to intervene, to set conditions, and so on."

REACH was finally approved in 2006 and entered into force in 2007. Nolan noted that it covers all aspects of all chemicals, including their manufacture, import, sales, and use, and it applies not just to chemicals on their own but also to chemicals in mixtures and in articles. Food, cosmetics, medicinal products, medical devices, and waste are specifically excluded from the regulation. The goals of REACH were to improve the health and safety of workers and the general public; to protect the environment by avoiding contamination of air, water, and soil and minimizing damage to biodiversity; and to maintain a competitive and innovative chemicals industry.

REACH was needed for a number of reasons, Nolan said. First, there was little or no information available on most of the chemicals in circulation. Second, the burden of proof was on the authorities, and the risk-assessment procedure was too slow. The downstream users were not involved, so the producers tended to know about the substances they were producing, but there was little information on the actual uses of the chemicals in products. Furthermore, the systems for dealing with the chemicals were inefficient, with a myriad of directives and regulations concerning chemical risk. REACH consolidated all those efforts and put them into one framework, Nolan noted. Finally, the administrative burden for new chemicals prevented innovation by imposing data requirements that did not apply to existing chemicals.

As an example of how downstream users may not be involved, Nolan discussed the situation with pesticides. There are a lot of bulk chemicals used in pesticides. These tend to be fairly simple molecules used in low volume by pesticide manufacturers, and they can be bought in bulk on the commodities market. The upstream producers of these chemicals—who may be manufacturing hundreds of thousands of tons per year of a particular chemical—have no interest in the low volumes being used in agriculture. As a result, Nolan said, "the users find themselves without the data and without the dossiers and basically without the possibility to keep using the chemicals because the production industry was not interested and maybe didn't even know these were being used for pesticides. So they came off of the market even though they may have been safe to use."

He also offered more detail on how the previous system prevented innovation. "We had treated existing chemicals and new chemicals differently. A lot of existing chemicals were grandfathered in, in our legislation, but we had higher data requirements for new chemicals. This prevented innovation in the chemicals market." Under REACH, all chemicals are being treated equally, and the hope is that this will promote innovation and substitution in the chemical industry.

REACH addresses each of the five shortcomings of the previous system that Nolan mentioned:

- To address the data gaps, a databank has been set up that is run by the European Chemicals Agency. "We make as much information public as possible and basically try to identify the gaps," Nolan said.
- Whereas the burden of proof used to be on public authorities, now it is on industry. "We save tax dollars and so on."
- In contrast to the previous system, which did not involve the downstream users, "now we are basically saying everybody should be involved," Nolan said. One problem, he explained, is that it is not always in the best interests of larger companies that can afford to generate the necessary data to share the data with smaller companies that cannot afford to generate the data on their own. "This is one of the biggest challenges that we have with REACH—how to get everybody to cooperate."
- To address the inefficiencies of the previous system, REACH incorporates and streamlines the legacy of earlier legislation.
- To lessen the administration burden for new chemicals that had been stifling innovation, there is no registration duty for low-volume chemicals.

A review of REACH was conducted in 2013, Nolan said, and one of its conclusions was that REACH is functioning quite well. "It is delivering on all of the objectives in the timeframe so far." There are a few adjustments that need to be made, so the legislation will need to be tweaked. However, he said, "given the preference of industry to have a stable regulatory environment, we are actually not planning to make any major legislative changes to it."

One of the findings of the review is that the regulations have had a big impact on small-to-medium enterprises. "A lot of them cannot access the dossiers," Nolan said. "A lot of them don't have the money to invest in dossiers." And even though a great deal of effort has been made in reaching out to these small manufacturers, many of them still do not even know about REACH. Thus, in the current framework there is a need to reduce the impact of REACH on these small-to-medium enterprises. The review also concluded that there are many opportunities for improvement by optimizing the program implementation at all levels and that the commitment of all actors involved is necessary.

Concerning REACH's effects on human health and the environment, Nolan said, the review concluded that it is still too early to quantify the benefits, but there are "positive trends in the sense that companies are looking for safer alternatives because they know there will be problems down the road when their substances start being reviewed." There has also been a noticeable move toward finding substitutes for substances of very high concern.

In January 2015 there will be a follow-up report on the effects of REACH on innovation, and it is expected that there will be Commission proposals for new data requirements concerning substances whose annual manufactured volumes fall between 1 and 10 tonnes (1,000 and 10,000 kilograms) and also possibly concerning polymers as well.

DISCUSSION

Dennis Devlin of ExxonMobil Corporation, the session moderator, opened the discussion session by asking Cleland-Hamnett whether the EPA and the nongovernmental organization community are making adequate use of the 10,000 technical dossiers on the European Commission's website. "I was involved with several of those dossiers, and I know they tend to be fairly extensive," he said. "I am hoping that information is being used beyond Europe as extensively as it can be."

Cleland-Hamnett answered, "We are making use of that information as much as we can. We look at it in terms of our prioritization process, and we have looked at it in terms of our risk assessment process." However, she said, one problem with the dossiers it that they don't provide access to the underlying studies. "When we are at the point of doing a risk assessment for a chemical, we can't really do that without having access to the underlying study and being able to look at those ourselves and then make it available to those who are peer reviewing and then looking at our risk assessments. I say that is the major limitation for us in terms of that data."

Denison agreed. The most disappointing thing about the implementation of REACH, he said, has been that it has not delivered the level of data and access to data that he and his colleagues had hoped for. "Generally speaking there are still a lot of unresolved issues around REACH and CBI protection generally that are being worked through the European court system and through the agency and industry negotiations," he said. "I think the jury is still out about the extent to which that will be the kind of source of information that we were all hoping on a global basis."

Nolan cautioned that people should keep in mind that REACH is still in its early days of implementation. "We are only 5 years into a process that could last 20 years and where it had been acknowledged at the outset that impacts could take 10 years or more to become evident," he said. "We are still at the organizational stages and putting the framework together. A new agency had to be set up and staffed with 500 people." He did acknowledge that there have been several problems, one of which is the varying quality of the dossiers that have been received. "The staff at the agency actually spent a lot of time chasing back to companies and saying, 'This isn't right, that is not right, this is missing, that is missing.' We really need to work a lot better with the companies to improve the quality of the dossiers." On the other hand, he said, big industry has bought into REACH. "It is supporting REACH, and it has more or less delivered what we would have wanted it to deliver." One problem does arise, however, with the industry consortia, in which users gather to agree on producing a common dossier. "It is only really the big companies that have the capacity to take the leadership of such a consortium," he said, "and many don't want to take the leadership."

Denison qualified his comments by saying that he is a big fan of REACH and that he was just pointing to one way it had fallen short of its goals. But, he said, the program has met a good number of its goals concerning the rate at which chemicals are being registered and the review of those chemicals. "They are on track to meet their 2018 deadline for registration of 30,000 substances, roughly," he said. "That is a remarkable achievement if that last leg happens."

Next, Jack Spengler, of Harvard University and a Roundtable member, offered an observation: "I think the market is not waiting for regulatory reform," he said. For example, the U.S. Green Building Council is revising its lead standards for buildings to have credit options that require the declarations of content of materials that might be specified in buildings. "I hope Harvard is joining the likes of Google and Kaiser Permanente and Stantech . . . in looking at these requirements for all of their new building specifications and even including red listed chemicals that they do not want in their products or in their building."

"There is no doubt that the market is responding," Walls responded. "We have seen several retailers, in addition to private standards developers like LEED [Leadership in Energy and Environmental Design] or U.S. Green Building Council, developing standards. . . . Walmart and Target have been developing one standard or another."

Cleland-Hamnett agreed, saying, "I think that we really do need to look to the market to really help us with this." It is quite a burden for the EPA alone to have to look at tens of thousands of chemicals, identify all of the uses and all of the potential risks, and decide what to take action on, she said. "I think it needs to be complemented by responsibility on the industry's part through the supply chain to understand what chemicals are in products and what we know and don't know about those products."

Faiyaz Bhojani from Shell Oil, a Roundtable member, asked about how the proposed revision of TSCA, the CSIA, differs from REACH. This is an important issue for international manufacturers such as Shell that must obey different sets of regulations in different countries.

Walls answered that the major difference between the two is that REACH has a mandatory minimum dataset, while the CSIA does not require a minimum dataset. "It does require that EPA get information sufficient to make a decision," he said, which "gives EPA a considerable amount of discretion in individual cases."

Another major difference is that the CSIA does not specify a list of substances of very high concern. REACH uses a completely hazard-based approach to identify chemicals of high concern, in which chemicals are listed because of their particular hazard characteristic. "Under the CSIA," Walls said, "EPA would identify particular priorities using a risk-based process and then would do a risk assessment for those substances and make some determination about their safety under the intended conditions of use. It is a different structure and a different legal standard as well."

Richard Denison added to the answer by pointing out a conceptual difference between REACH and the CSIA. "The primary thrust of REACH was this concept of shifting the burden of proof from government to show harm to industry to show safety," he said. "What it actually means under REACH is that the industry does the assessments and decides what risk management is necessary and communicates that risk management through the supply chain. That model only works if people have confidence in that information and the government's ability to make sure that is accurate."

By contrast, under the CSIA it is government that does the assessments and decides on risk management. "That is fundamentally a different concept," Denison said. On the other hand, TSCA reform will lead to a shift from a presumption of safety to an affirmative requirement that safety be demonstrated, bringing that more in line with REACH. The difference will be in who is responsible for proving safety.

Liz Harriman with the Massachusetts Toxics Use Production Institute asked about the balance between the states and federal government in assessing and regulating chemicals. In particular, she asked, "Can you say a little bit about your various thoughts on the provisions under the TSCA bill for preemption of states and how you see that balance between preserving states rights to act versus creating a level consistent message for companies at the federal level?"

Walls answered the United States needs a robust uniform national system of chemical regulation. "We set safety standards for automobiles at the federal level, we set safety standards for pharmaceuticals at the federal level, and there is still in both cases room for the states to act appropriately," he said. "Even under TSCA today there is a waiver provision that allows states to seek a waiver from the preemptive effective of current TSCA decisions. It has never been used in 37 years."

In short, Walls said, "I would like to see predominantly a federal system." He said he believes that the CSIA strikes the right balance. "Let's engage the states in the discussion around risk assessment and potential risk management actions so that we can make sure the decisions that are made are protective across all 50 states."

Denison agreed with Walls that a balance needs to be struck. "However," he said, "I do think the current bill is out of balance. It preempts state authority to too excessive a degree." He clarified that he does believe there is a need for a strong national program. "We need a program that raises the floor nationally, but we need to make sure that states still have the authority to act where the federal government hasn't, and until the federal government does. We also need a waiver provision that is workable that allows states to act when there are reasons for them to do so even where the federal government has acted."

"The biggest concern I have about the bill," Denison continued, "is that it would preempt new state actions merely on the basis of a prioritization decision by EPA that a chemical is either low or high priority. In particular, if it is a low-priority designation, that action is not even judicially challengeable. It is effectively a final agency action. It would have a preemptive effect on states, and I think that is a toxic combination. While I don't disagree that one of the big drivers that got the industry to come to the table is the need for a national program, we don't have the balance right in that bill yet."

David Andrews of the Environmental Working Group noted that Canice Nolan had said in his presentation that one of the motives for instituting REACH was worries that the previous regime may have been stifling innovation because of higher data requirements for new chemicals compared to those for existing chemicals. Thus, he wondered if, because there are different standards for new and existing chemicals under TSCA, this situation might be doing something similar in the United States. "There are really no regulations on existing chemicals and no new test orders, while for new chemicals, there are an increasing number of SNURs on new chemicals and increasing review by EPA," he said. Is that stifling innovation?

Walls answered that it depends on the particular SNUR on whether it stifles innovation or not. "There have been SNURs that have basically halted all further development in those areas," he said, although the evidence is circumstantial. In general, he said, the question of how REACH has affected innovation is still very much unanswered. "The Commission is still conducting a review," Walls said. "There is some anecdotal evidence. We still have around three times more new chemicals being introduced in the United States than in any other region. The different regulatory structures could be the major factor. I think it is still too early to tell what impact REACH will have."

Cleland-Hamnett clarified that what a significant new use rule requires is that a company that wants to pursue an activity that is identified as a significant new use needs to submit a notice like a premanufacture notice. "The SNURs in and of themselves don't impose regulatory requirements beyond what the first manufacturer or importer had to do to get that chemical on the market. It is really leveling the playing field more than anything else. Because of the way TSCA is set up, the first company coming through the gate, we can ask for testing and put certain limitations on the use of the chemical pending the development of the testing." By contrast, any company that follows that first company in making or using that same chemical will not be subjected to those requirements. "If you look at the innovation being the creation and the application of the chemical to begin with, leveling the playing field to additional companies that want to piggyback on using that chemical is a positive step."

Denison said that it is important to keep in mind that REACH did not address the disparity between new and existing chemicals by lowering the bar for new chemicals. It did it by raising the bar for existing chemicals. He would prefer a system that essentially requires a universal scenario, he said. "Wherever EPA looks at a chemical and decides that it is safe under a set of conditions, those conditions [should] be encoded in some way that others would then need to comply with. If they wanted to depart from those, they would need to go through a process of ensuring the continued safety of that chemical." One problem, he said, is that SNURs have been used as a way to deal with bad chemicals. "I think a more universal approach would be preferable."

Walls responded to that by saying that the problem with a blanket SNUR is that does inhibit innovation. "Most of the innovation and chemistry is coming from chemicals that are already on the market."

Andrew Maguire, a Roundtable member, asked exactly where in the process the proposed legislation to reform TSCA is now. Christina Franz, Senior Director of Regulatory and Technical Affairs at the ACC, answered that a House hearing was scheduled for the following week, on November 13.[7] "I don't believe the witness list has been noticed yet," she said, "but the House is convening a hearing on the CSIA, which is the Senate bill. That, in and of itself, is a significant development because it is highly unusual, if not unprecedented, for the House to entertain a hearing on a Senate bill. It is worth watching, I think."

There have been reports that the two sponsors of the bill, Senators Vitter and Udall, are working with and speaking with nongovernmental organizations and other stakeholders who have expressed concerns over several areas of the CSIA and, apparently, are trying to work toward making accommodations going forward. Franz noted that "we haven't seen any amendment or revision of the CSIA at this juncture."

Franz said she is hopeful the bill will pass. It is a bipartisan bill. "It has now, I think, 26 signatories, half Republican, half Democrat," she

---

[7] The U.S. House of Representatives Committee on Energy and Commerce, Subcommittee on Environment and the Economy held a hearing on November 13, 2013, titled "S. 1009, The Chemical Safety Improvement Act." The witness list, written testimony, and archived video are available at http://energycommerce. house.gov/hearing/s-1009-chemical-safety-improvement-act (accessed March 31, 2014).

said. "That is a huge achievement. That, certainly, should give us all optimism, particularly in this Congress, that something should be able to be accomplished." In particular, she said, the bipartisan nature of the bill separates it from previous efforts to reform TSCA, none of which were bipartisan. "The fact that the CSIA is a bipartisan bill is a huge change in the landscape."

Gina Solomon of the California Environmental Protection Agency (California EPA) added that, to her, there are some aspects of the CSIA bill that are really very promising. For example, it would give the EPA the ability to more easily share chemical information with the states. "Unfortunately," she said, "it also grandfathers in old CBI claims, which operates in the contrary direction."

The California EPA Department of Toxic Substances Control put a lot of thought into the issue, she said, and it settled on three principles to guide its policies on CBI in the California Safer Consumer Product regulations. First, there should be very significant substantiation of CBI claims, including showing that the information is very guarded in other contexts and that the information cannot readily be reverse engineered through analytical chemistry or other approaches. Second, those claims should be revisited periodically "because something that is bona fide CBI today may not be a year from now." Third, issues of CBI should not be connected to hazard trait information. "In other words, if you are presenting toxicology data or other hazard data on a chemical, you have to show that a patent has been applied for and is still pending, at which point you could temporarily mask the chemical identity associated with the hazard information until the patent is approved." In short, she said, "you can put some constraints around CBI and still maintain it. That is at least what we are hoping to do in California."

Goldman echoed and expanded on Solomon's remarks on CBI. "I think that your point is well taken that CBI needs to be revisited," she said. "I would even go further and say claims perhaps could have a sunset time and that, if companies want them extended, they would have to justify that. But I would guess that most of the claims could be sunsetted with no harm to the industry."

It has been far too easy for companies to claim information as CBI, Goldman said. "When we did an audit years ago at EPA, the auditors found a *New York Times* article stamped CBI." The problem is that there has been no question of whether things are actually CBI. She explained that if you work for a company and mistakenly disclose information that should have been stamped as CBI, there could be trouble. If you over-claim, there likely is no trouble because nobody has been checking and no real penalty is in place. "One problem is that the EPA's CBI systems are based on old paper type systems," she said. "If the systems were updated and modernized using modern informatics, it would be easier to implement reforms."

Goldman suggested that, as is the case for pesticide registration, the companies be required to submit all the raw data from their studies of industrial chemicals. The EPA would have no need to disclose all this information to the public, and most members of the public do not really want to see all the data, she said. However, in the case of pesticides, the EPA does disclose summaries of the studies to show the basis for safety determinations. "That does not get the companies into trouble with CBI because data summaries cannot be taken to a regulatory authority in Argentina, Brazil, or somewhere else to get a pesticide registration," she said. "This is a commonsense way that the hazard information can be disclosed without hurting somebody's investment."

The bill also gives the EPA some very sweeping preemption powers, Solomon said, far more sweeping than in existing TSCA, and these provisions are really worrisome to California and to other states in their current form. "In particular, preemption kicks in long before EPA actually promulgates any regulation," she said. "As soon as a chemical is prioritized, then some preemption kicks in. It could be years or decades until EPA actually takes action. During that time, states would not be able to take action, which is unfortunate."

Furthermore, as Denison pointed out, states would have no ability to act on anything that the EPA designates as low priority. "There are chemicals, for example, that would be of low toxicity, but high concern to a state like California because of their high global warming potential," Solomon said. "Yet, despite a provision that says we could act using air and water quality laws, we really can't because there is also a provision that says we can't do anything if it interferes with the production, distribution, or use in commerce of that chemical."

REFERENCES

EEA (European Environment Agency). 2010. *The European environment—state and outlook 2010: Synthesis*. Copenhagen: EEA.

EPA (U.S. Environmental Protection Agency). 2009. *EPA announces actions to address chemicals of concern, including phthalates: Agency continues efforts to work for comprehensive reform of toxic substance laws.* Available at http://yosemite.epa.gov/OPA/ADMPRESS.NSF/d0cf6618525a9efb85257 359003fb69d/2852c60dc0f65c688525769c0068b219!OpenDocument (accessed March 31, 2014).

EPA. 2012. *High Production Volume Information System (HPVIS).* Available at http://www.epa.gov/hpvis/ (accessed March 31, 2014).

GAO (U.S. Government Accountability Office). 2013. *Toxic substances: EPA has increased efforts to assess and control chemicals but could strengthen its approach.* Washington, DC: GAO.

4

Models for Environmental Risk Assessment
and Exposure Science

After two sessions devoted to providing background and context to the topic of chemical risks and the assessment of those risks, the workshop's third session examined some of the seminal work in the field that has been done or has been in progress over the past few years. In particular, the third session's speakers described two recent major reports on the subject, *Science and Decisions: Advancing Risk Assessment* (NRC, 2009) and *Exposure Science in the 21st Century: A Vision and a Strategy* (NRC, 2012); a conference held by the U.S. Environmental Protection Agency (EPA) in 2011 and subsequent report on the same topic titled *Next Generation of Risk Assessment: Incorporation of Recent Advances in Molecular, Computational, and Systems Biology* (EPA, 2013); and a current project being conducted by the National Research Council (NRC), the Design and Evaluation of Safer Chemical Substitutions: A Framework to Inform Government and Industry Decisions, which is intended to produce a consensus report that will be released in the fall of 2014.

SCIENCE AND DECISIONS: ADVANCING RISK ASSESSMENT

In the first presentation John Balbus, Senior Advisor for Public Health at the National Institute for Environmental Health Sciences, described the 2009 NRC report *Science and Decisions: Advancing Risk Assessment*. He served as a member of the committee that wrote the report.

The subject of that report—risk assessment—is just one part of the topic of the current workshop, he noted. "The regulation of chemicals involves a lot of different processes, some of which are steps in risk assessment," he said. "A lot of what goes on under TSCA [the Toxic Substances Control Act of 1976[1]] is really just hazard identification because we don't have the kind of robust information that is required in a risk assessment."

So one question that must be asked is, How does one gather exposure assessment information that can help in the prioritization of regulatory decisions when there is not enough information for a full risk assessment? "I think that this [2009] report has some thoughts on that that is relevant here," he said. Of course, he added, there are also situations in which there is plenty of information for making such regulatory decisions, and it is important to recognize the difference in approach when dealing with an information-poor environment versus dealing with an information-rich environment.

The charge to the committee that produced the 2009 report was to

- develop scientific and technical recommendations for improving risk analysis approaches used by the EPA, including practical improvements that the EPA could make in the near term (2–5 years) and in the longer term (10–20 years) (NRC, 2009) and
- focus primarily on human health risk assessment, but also consider broad implications of findings and recommendations for ecologic risk analysis (NRC, 2009).

Balbus emphasized that the committee was looking not just at short-term, very practical recommendations but also at long-term, aspirational recommendations. He noted that the committee asked, if we could really do this the way we wanted to, what would be the goal to set and how would we advance the science to get there?

The members of committee also decided early on that they would look at two different elements of risk assessment. "One was the technical side, the nuts and bolts, the science," he said. "How do you conduct the stages of the risk assessment, and how can we improve that technical conduct?" The second elements was on the decision-making side, with a focus on how to increase the utility of risk assessment. "How can we alter the framework and the ways that we think about risk assessment, and how do we do risk assessment in a way that improves its utility in the kind of decisions that have to be made?"

Balbus presented a list of the key messages from the report that he would further explain: (1) enhanced framework, (2) formative focus, (3)

---

[1] Toxic Substances Control Act of 1976, Public Law 94-469, 94th Congress.

four steps still core, (4) matching analysis to decisions, (5) clearer estimates of population risk, (6) advancing cumulative assessments, and (7) people and capacity building. These were the official take-home messages determined by the committee, Balbus noted. "A lot of the discussion was about the framework for decision making and how risk assessment plays a role in it," he said, referring to the first bullet point. "'Formative focus' means there was a lot of focus on the setup, the questions asked at the beginning." In particular, "formative" refers to the process of forming the questions for the risk assessment. The third bullet point refers to the fact that the four steps from *Risk Assessment in the Federal Government: Managing the Process* (NRC, 1983), known as the Red Book, are still accepted as the key to the risk-assessment process.

"There was a lot of discussion about 'right-sizing' risk assessment because risk assessment can get very complex, very involved, very expensive, and very lengthy," he said. "How do we do that at the right times and the right places but at the same time figure out ways to do good, but less involved, risk assessments where the decisions are better served by that kind of an analysis?" The final bullet point referred to the committee's belief that there is a need for awareness raising and training, both among risk assessors and risk managers, as well as a need for capacity building.

Balbus then returned to the fifth and sixth bullet points to deal with them in more detail, as he said he felt they were the most relevant to the topics of the workshop.

To obtain clearer estimates of population risk—the fifth bullet point—it will be necessary to deal more effectively with uncertainty and variability in those estimates. One of the report's recommendations was that the EPA "should encourage risk assessments to characterize and communicate uncertainty and variability in all key computational steps—for example, exposure assessment and dose–response assessment." In particular, Balbus said, the report recommended that "uncertainty and variability analysis should be planned and managed to reflect the needs for comparative evaluation of the risk management options."

There was a great deal of discussion in the committee concerning how to determine the right amount of uncertainty analysis for the particular kind of assessment being undertaken, Balbus said. The recommendation was that, in the short term, the EPA "should adopt a 'tiered' approach for selecting the level of detail to be used in the uncertainty and variability assessments, and this should be made explicit in the planning stage."

One of the notable points to come out of the committee's discussions, Balbus said, was that although there is generally a great deal of attention paid to the uncertainty and variability in the toxicology and dose response of a particular substance, much less attention is paid to the uncertainty and variability on the exposure side.

A second area that received a lot of discussion was the selection and use of defaults, he said. "The overarching theme here was to call for much more transparency and much more explicit discussion of defaults and their basis." But a secondary theme that emerged was all the implicit defaults that are used, sometimes unconsciously, in risk assessment. "There are a lot of assumptions that are inherent in the way we do risk assessment that are never called out as being defaults," he said. "A lot of the discussion in the committee was about recognizing these and identifying these and maybe thinking about them a little bit differently."

A key default appears in site risk assessments where there are multiple different chemicals to which people are exposed. "If a particular chemical doesn't have sufficient information, it is by default assumed to have zero risk," Balbus said. "This isn't considered in the risk assessment. You do the risk assessment for the chemicals that you know about, and anything else that is in there [is assumed to have] zero risk. It may be a good default, or maybe there should be a default that if you don't anything, it has some kind of average risk [for that particular class of chemicals]."

A second implicit default is that carcinogens have a linear dose response and, furthermore, that they do not have any human individual variability. "These are the kinds of things that were brought up by the committee," he said. "I think this has some relevance to the way we frame risk assessments and even do some of the hazard identification."

Another issue that is relevant to risk assessment, Balbus said, is the criteria used to decide when not to use a default assumption. "The committee determined that EPA, for the most part, has not yet published clear, general guidance on what level of evidence is needed to justify use of agent-specific data and not resort to a default." Among those who carry out risk assessments, "there may be different criteria used for when you depart from a default assumption," he said. "So there was a call to provide guidance and have transparency and clarity about this." The committee noted that there are also a number of defaults that are engrained in the EPA risk-assessment practice but that are absent from its risk-assessment guidelines, Balbus said. With respect to the selection and use of defaults, the committee made three recommendations in the report:

- "EPA should continue and expand use of the best, most current science to support and revise default assumptions.
- EPA should work toward the development of explicitly stated defaults to take the place of implicit defaults.
- EPA should develop clear, general standards for the level of evidence needed to justify the use of alternative assumptions in place of defaults" (NRC, 2009).

One of the most controversial parts of the committee's work was its recommendations for how the EPA should unify its approach to dose-response assessment for both carcinogens and noncarcinogenic substances. The committee thought that the EPA's treatment of non-cancer and low-dose, nonlinear cancer end points is a major step in an overall strategy to harmonize cancer and noncancer approaches, Balbus said, but the committee also found that there are scientific and operational limitations to this approach.

In particular, the committee focused on the issue of thresholds. While there may be a "clear red-line threshold" for a given effect and a given substance when considering a particular individual, "that threshold disappears when you consider it a population level," Balbus said. "The committee did a lot of thinking and a lot of deliberation on what might be the approach by which you would unify dose response for carcinogens and noncarcinogens to take into account this population-scale lack of a clear, defined threshold." The committee's approach to unification involved "investigating this interindividual variability in susceptibility, looking at population susceptibility and at the influence of underlying disease on vulnerability and looking at other kinds of genetic susceptibility factors, quantifying that, and then taking a look at mode-of-action information and other toxicological information." The approach that the committee developed is laid out in Figure 4-1.

The first step, Balbus explained, is to look at the toxicological data and at the end points and the nature of those end points and to understand, to the extent possible, the biological mechanisms of those end points. "The second tier is where you start looking at mode of action, but at the same time you bring in consideration of vulnerability factors and the distribution of those vulnerability factors and the importance of them as the variability in background exposure. Then you can use informed expert judgment to decide upon the right conceptual model. In many cases the committee believed that this would lead to a more linear approach to most impacts without clear bright-line thresholds."

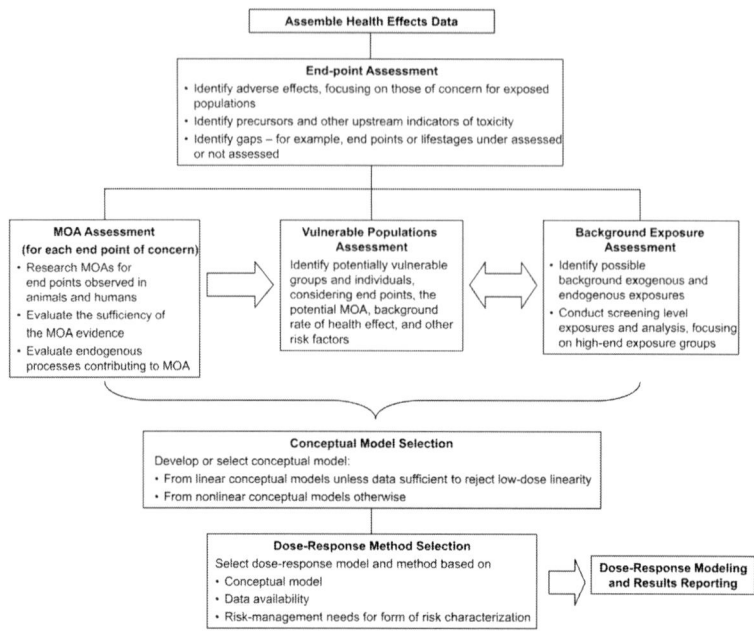

FIGURE 4-1 New unified process for selecting approach and methods for dose–response assessment for cancer and noncancer end points.
NOTE: MOA = mode of action.
SOURCE: NRC, 2009.

Another important issue that the committee focused on in the report is cumulative risk assessment. There is a growing realization among environmental health professionals, Balbus said, that it is not sufficient to examine single-agent exposures in a vacuum. "You have to look at exposures in context—not only in the context of co-exposures with other chemicals, but also in a context of multiple nonchemical stressors, whether that is psychological stress, nutritional stress, or socioeconomic stress." The committee recommended that the EPA should "draw on other approaches, such as those from ecological risk assessment and social epidemiology, to incorporate interactions between chemical and nonchemical stressors in assessments." In the short term, the EPA should "develop databases and default approaches to allow for incorporation of key nonchemical stressors in cumulative risk assessments in the absence of population-specific data, considering exposure patterns, contributions to relevant background processes, and interactions with chemical stressors."

In the longer term, the agency should invest in research programs related to interactions between chemical and nonchemical stressors, including epidemiologic investigations and physiologically based pharmacokinetic modeling."

In summary, Balbus reiterated that there were two different aspects to the committee's recommendations. The first was the technical side—various ways to improve the validity and usefulness of the assessments. "The other piece was imbedding risk assessment in a decision framework that considers the question that has to be answered and the choices that have to be made," he said. "For example, in green chemistry if you have a multitude of different kinds of chemicals that you could be using other than your chemical of concern, that would be a very different risk assessment than if you only have that one chemical to serve a particular function." Thus, the committee recommended that the EPA "adopt a framework for risk-based decision making that embeds the Red Book risk-assessment paradigm into a process with initial problem formulation and scoping, up front identification of risk-management options, and use of risk assessment to discriminate among these options."

EXPOSURE SCIENCE IN THE 21ST CENTURY:
A VISION AND A STRATEGY

The second speaker was Paul Gilman, senior vice president and chief sustainability officer at Covantra. He served on the National Academy of Sciences (NAS) committee that produced the 2012 report *Exposure Science in the 21st Century: A Vision and a Strategy* (NRC, 2012).

Gilman began his presentation with a question: "If you are a toxicologist or a risk assessor or an epidemiologist, you have an appreciation for exposure science. In fact you might even say some of your best friends are exposure scientists. But do you ever really invite them to the buffet?" Gilman said that the purpose of his talk would be to argue that "they should be at the buffet table and be a critical component."

Before discussing the 2012 report, Gilman offered an anecdote to emphasize the importance of exposure science. In the year following the September 11, 2001, attacks, a group at the EPA put together a screening tool for examining some of the potential hazards the United States faced. The group took a unique approach in that a significant component of the screening tool considered exposure and exposure scenarios. "On the basis of that work they ran tens of thousands of scenarios," he said. They examined a number of scenarios on a classified list that was circulating in government and came up with a modified list that was dramatically different. "Using exposure they could strike lines through a number of the scenarios that were on that list and introduce new ones that probably merited higher consideration and preparation," Gilman said. "It was sufficiently novel that all of a sudden the nascent Department of Homeland Security wanted to understand it. People at the White House wanted to understand it. For many of them who were from the toxicology side of things, it was eye opening. That was in a way my own grounding in the importance of exposure to the consideration of significant real-world problems."

The goal of the study, Gilman said, was to provide guidance on how to use exposure science in the regulatory arena. As part of that the committee developed a conceptual framework showing the core elements in exposure science and how they are linked (see Figure 4-2).

One of the main focuses of the study was the various scientific and technological advances that have emerged in recent years that can be applied to exposure science. These include geographic information technology, ubiquitous sensing techniques, the use of biomonitoring for assessing internal exposures, and the modeling and information-management tools. Another tool that is still emerging is the use of crowd-sourced information on the exposure side. To illustrate the roles of these new tools in exposure science, the committee created a modified conceptual framework (see Figure 4-3).

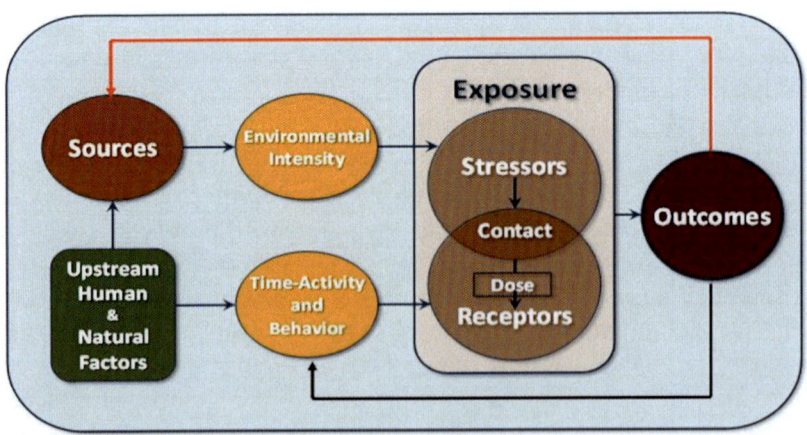

FIGURE 4-2 Core elements of exposure science.
SOURCE: NRC, 2012.

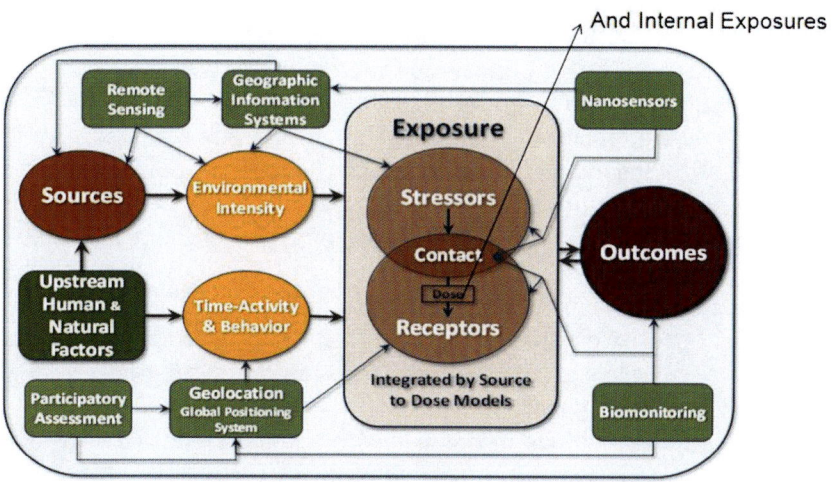

FIGURE 4-3 Expanded view of the core elements of exposure science.
SOURCE: Adapted from NRC, 2012.

These emerging technologies combined with a growing appreciation for the power of exposure analysis make exposure science "a place that we might think of as an emerging frontier and one that should be focused on," Gilman said. The committee laid out a vision of exposure science that is moving from the historical focus on discrete exposure to a new, broader focus that considers exposures

- "from source to dose;
- on multiple levels of integration (including time, space, and biologic scale);
- to multiple stressors; and
- scaled from molecular systems to individuals, populations, and ecosystems" (NRC, 2012).

To capture that new, broader conception of exposure, the committee relied on the notion of an "eco-exposome." The key idea behind this notion, Gilman said, is that exposure can no longer be thought of in terms of a single stressor occurring at a single point in time. Instead exposure should be conceived of in developmental terms. "It changes through time. It is influenced by all of the other stressors affecting the organism."

It is a powerful, all-encompassing vision of exposure, Gilman said. "The good news is in thinking about it you have to know everything about everything all of the time." On the other hand, it is a vision that seems to require quite a lot to fulfill. "The bad news is you have to know everything about everything all of the time." And so in an era of increasingly tighter budgets for research, which are forcing agencies to be increasingly careful in prioritizing the research that they fund, this conception of exposure science can make the field difficult to "sell" to funders, Gilman said. "People would say that it is just too hard. It is too much. It costs too much. It is a 20-year program of trying to know everything about everything for an organism and all of the organisms that affected it because certainly we were stressing putting people in an ecological context as well."

With those difficulties in mind, the committee identified two overarching research needs in the area of exposure science:

- "Characterizing exposures quickly and cost effectively at multiple levels of integration—including time, space, and biologic scales—and for multiple and cumulative stressors, and
- Scaling up methods and techniques to detect exposure in large human and ecologic populations of concern" (NRC, 2012).

"This is again the notion of needing to know everything about everything for all time when considering a stressor and the stressed organism or organ or cell," Gilman said. "But while the advent of new sensing technologies and approaches to looking at information can allow this, you have to try to come to grips with it in a way that you can make progress and not just wait until you know everything about everything all of the time."

The committee's strategy for meeting these research needs was to focus on the urgent needs of the day and to use those urgent needs to develop the tools and the infrastructure for carrying out the research. That infrastructure includes the educational infrastructure that will be needed to train researchers in new approaches and to teach them how to integrate these different tools to answer specific questions. "Then using this infrastructure and using these tools, you can begin to look to those environmental health–related hypotheses that are the more general questions," Gilman said. "They are the questions that will lead us to a point when we can look at exposure science as a predictive science."

One thing that could help push that strategy along much faster would be improved collaboration among the various institutions that have knowledge and capabilities relevant to exposure science, he said. "There are so many places in our federal research agencies, in our private research institutions, and in our public research institutions that have information that is . . . in one place and not shared. It is certainly not integrated." The 2012 report "is rich with ideas about experiments that could be done, monitoring programs that could be done on very large scales and across very diverse places, looking at all organisms through all stages of their life with all of the stressors," he said.

ADVANCING THE NEXT GENERATION OF RISK ASSESSMENT

The next speaker, Ila Cote, a senior science advisor for the EPA's National Center for Environmental Assessment, described a conference, Advancing the Next Generation of Risk Assessment (EPA, 2011a), and a subsequent report on the same topic (EPA, 2013).

The Next Generation of Risk Assessment, or NexGen for short, is an effort that has been going on at the EPA for almost 4 years.[2] NexGen is looking at if it should and how to use recent advances in molecular, computational, and systems biology to better inform risk assessment, Cote said. The EPA is joined in the effort by a number of partners, including the National Institute of Environmental Health Science, the National Toxicology Program, the Department of Defense Army Corps of Engineers, the Food and Drug Administration National Center for Toxicologic Research, the National Institute of Occupational Safety and Health, Health Canada, the California Environmental Protection Agency, the European Chemical Agency, and the European Community Joint Research Commission.

The NexGen effort started with a review of the recommendations in several earlier reports, including *Toxicity Testing in the 21st Century* (NRC, 2007), *Science and Decisions* (NRC, 2009), and *Strategic Plan for the Future of Toxicity Testing and Risk Assessment at EPA* (EPA, 2009), as well as information presented in workshops from the NRC's Standing Committee on Emerging Sciences for Environmental Health Decisions.[3] One of the recommendations common to the reports was to

---

[2] Further information on NexGen is available at http://www.epa.gov/risk/nexgen (accessed April 2, 2014).

[3] Further information on workshops from the Committee on Emerging Sciences for Environmental Health Decisions is available at http://nas-sites.org/emerging science (accessed April 2, 2014).

develop case studies or prototypes that provide concrete examples of new types of risk assessments and "engender movement from strategy to practical application of new assessment approaches," Cote said. So development of prototype risk assessment was an important focus of the project. Seven different prototypes illustrating the use of different types of molecular, computational, and systems biology data were developed.

Several other activities were undertaken in preparation for prototypes development:

- Meetings were held with decision makers to learn about their information needs that might be met by new risk-assessment approaches. Matching analysis to decisions is one of the key messages from the NAS, as noted earlier by Balbus. The risk-assessment process can be more efficient if "fit-for-purpose" or "right-sized" assessments are developed, she said. Hence, from discussions with decision makers and reviews of the available data, prototype concepts were developed that could support different decision contexts.
- A draft strategic framework that articulated guiding principles for NexGen prototypes development was also developed.
- A meeting was held, November 2010, for experts to discuss the framework and help refine the prototype concepts (EPA, 2011b).
- A public meeting was held in February 2011 to communicate the intended process and to take comment on the plan from a diverse group of stakeholders, as well as to gather advice on how to communicate during the process and the results of the effort (EPA, 2011a). "To some extent the final report is a fulfillment of promises that we made at this public dialogue conference," she said.

One of the things that came out of the public dialogue conference was a better understanding of the kinds of information that stakeholders want, Cote said. "[EPA] decided that there was a need for a series of technical papers that were primarily targeted to the scientific community. There were approximately 40 papers that either have already been published or will soon be published that are products of NextGen." The project also produced a summary report of the technical papers, and an executive summary targeted at risk managers and the lay public. The draft report has completed external peer review and public comment and will be final in spring 2014.[4]

Cote said for illustrative purposes the EPA tried to look broadly at three kinds of decision-making situations and develop three categories of assessments that could address those needs. The three categories are major-scope assessments, limited-scope assessments, and prioritization and screening. As you move from the first category to the last, Cote said, the number of chemicals that a decision maker has to consider grows sharply from a few hundred to thousands to tens of thousands. Concomitantly, the amount of traditional data available to support decisions declines as the numbers of chemicals increase.

An example of major scope assessments, she explained, is something the agency is familiar with—the EPA's Integrated Risk Information System (IRIS) assessments or the Integrated Science assessments—"where one is dealing with relatively few chemicals with lots and lots of data." The EPA uses this type of situation primarily to develop proof-of-concept prototypes, and secondarily to explore how already robust traditional risk assessment might be better informed by new types of data. An example of a limited-scope assessment might be something like dealing with Superfund site cleanup or an emergency response where you might have a few thousand chemicals that a risk manager might have to consider. Finally, an example of prioritization and screening would be to rank "potentially tens of thousands of chemicals" in the environment for additional research, testing, or assessment. A second example of prioritization and screening is choosing safer and sustainable chemicals for use in the society.

Cote then focused the remainder of her talk on the first of the three decision context/assessment categories and described the major-scope assessment prototypes that have been completed. These prototypes are the proof-of-concept assessments that focused on "reverse engineering" from known public health risks to NexGen-type risk assessments, thus verifying the use of new approaches by comparison with robust traditional data. The first three prototypes examined the connections among benzene, other leukemogens, and leukemia; between ozone and lung inflammation; and between polycyclic aromatic hydrocarbons and lung cancer. These are all areas in which a great deal is already known, Cote noted, and that was exactly the point—to use robust traditional datasets to verify how to

---

[4] The public comment deadline was extended to January 13, 2014, and the final report will likely be available in fall 2014.

best use new data types. These proof-of-concept prototypes focused on evaluation of invivo human exposures (molecular epidemiology or clinical) at environmental concentrations where traditional and molecular data was collected concomitantly, and exposure–dose relationships were measured. "There are not many datasets that are like that," she said.

Cote noted that what the NexGen project intended to do was iterate back and forth between the new types of data, such as omic data and cell biology data, and the traditional data to understand what could and could not be done, sort out what information was most valuable, and begin to develop decision rules that would help the EPA use new types of data consistently and appropriately to get the "right answer." Several important points were illustrated by these major assessment prototypes, said Cote.

1. The molecular epidemiology and molecular clinical studies demonstrated that it is possible to identify molecular patterns that are predictive of specific hazards (i.e., disorder and disease) and exposure–dose responses, Cote said.

2. Chemicals that induce the same health outcomes appeared to share mechanistic commonalities. This is important, she said, because identifying underlying molecular patterns of disease could help characterize chemicals for which few traditional data are available but for which molecular mechanism or adverse outcome pathways are characterized. If, for example, a chemical is known, through in vitro data, to similarly affect the same genes and pathways affected by chemicals known to cause leukemia, it would be reasonable to assume that the relatively unstudied chemical might increase the risks for leukemia as well. Thus, understanding disease mechanisms or key steps in mechanisms could help the EPA evaluate large numbers of chemicals without traditional data.

3. Chemically induced mechanisms of disease appear to have many commonalities with naturally occurring disease or diseases of unknown origins. This will allow the EPA to utilize the large amount of mechanistic information on disease that has been developed for clinical reasons by the National Institutes of Health and others to help understand environmentally induced alterations in health.

4. Disease mechanisms do not parse cleanly into cancer and noncancer mechanisms. For example, pathways involved in altered

immune responses, inflammation, cell repair, and apoptosis contribute to cancer and noncancer health outcomes. Thus, new methodologies will require harmonized approaches to cancer and noncancer end points.

5. With new higher-throughput methodologies, it is possible to collect experimental data over a wide variety of potential exposure concentrations and hence refine or replace inferences about low exposure–response relationships with experimental data.

6. It is also easy to see how information on the impacts of various chemicals on the same mechanistic pathways could be used to evaluate the cumulative risk of mixtures of chemicals, Cote said. "Obviously, chemicals that interact with the same pathways are more likely to interact in terms cumulative risk than chemicals that don't interact in the same pathways." Chemicals and nonchemical stressors could be evaluated via their pathway interactions.

7. Lastly, in the prototypes it was clear that variations in human genes can alter responses in subpopulations. New approaches can help the EPA better characterize variability in overall population responses to chemicals, as well as less sensitive and more sensitive subpopulations.

Cote closed by discussing that what was learned from the NexGen prototypes will inform how the EPA will move forward to improve risk assessment. Of the key issues raised in *Toxicity Testing in the 21st Century* and *Science and Decisions*, the NexGen report (EPA, 2013) discusses how the agency might proceed on a number of issues, including matching assessments to decision context, harmonization of cancer and noncancer approaches, better characterization of population variability, cumulative risks from mixtures and other environmental stressors, and improved assessment of responses at environmental exposure levels.

She also discussed what has been learned from the NexGen project with respect to the significant challenges facing those who wish to improve risk assessment. First, a great deal of unusable data exists. "If we had known then what we know now, a lot of the studies we have would not have done in the same way," she said. Consequently, "systematic review and adherence to best practices for data used in risk assessment is going to be critical." A second challenge will be consideration of variability in the data. "The signal-to-noise ratio can eat you alive in these studies," Cote said. Consequently, characterization of variability is very important. Third, a whole new set of uncertainties will need to be described—new uncertainties "that we haven't spent the last 30 years discussing," she said. Fourth, it will be critical that the molecular changes under study be imbedded in a mechanistic network or context. "To be able to separate out what is just normal biology from disease biology requires that you put these things in more of a network context." Lastly, she noted that concomitant improvements in exposure science are needed. In Particular, easily measured biomarkers of exposure and effect are becoming possible, making possible direct measurement in exposed populations of altered biology and potential adverse health outcomes.

Looking to the future, Cote said, it will be important to develop an integrated understanding of cell biology. "One of the things you don't often see is studies that look at multiple biologic processes measured by various omic and cell biology techniques. For example, people tend to look at proteomics or genomics or transcriptomics but not at the integrated activity of these processes. You don't see these kinds of tools brought together in single sets of experiments." Enough is known to understand that many things are going on with chemically induced alterations in biology. "I would like to advocate for studies that take a more integrated approach using a variety of new methodologies."

Finally, she said, it will be necessary to start developing a dynamic—as opposed to a static—understanding of what happens in response to chemical exposures. "The studies we tend to have are snapshots in time of evolving biologic events. My colleague Lyle Burgoon says science currently gives you a roadmap of disease processes, but what you are really interested in is traffic flow." In other words, "we are really interested in information flow in the organism," rather than events at a point in time. However, she added, "we are not there yet" in terms of our analytic tools.

In conclusion, she suggested that the most promising approach to risk assessment will be to collaborate across different fields of study. "I think that integrating what we know about human disease and human genetics that comes out of the study of disease in the absence of chemical exposure with data on the effects of chemical exposures . . . is going to give us the best information that will allow us to screen new chemicals more rapidly using new molecular methodologies."

THE DESIGN AND EVALUATION OF SAFER CHEMICAL SUBSTITUTIONS: A FRAMEWORK TO INFORM GOVERNMENT AND INDUSTRY DECISIONS

The session's final speaker was Marilee Shelton-Davenport, a senior program officer with the Board of Life Sciences for the NRC of the NAS. She is the study director for a current NRC study, the Design and Evaluation of Safer Chemical Substitutions: A Framework to Inform Government and Industry Decisions.[5] That study will result in a consensus report, which is scheduled to be released in the fall of 2014.

Shelton-Davenport began by discussing a workshop that preceded her study, Applying 21st Century Toxicology to Green Chemical and Material Design, which was held in Washington, DC, in September 2011. It was hosted by the Committee for Emerging Science for Environmental Health Decisions at the NAS and sponsored by the National Institute of Environmental Health Sciences (NIEHS).

"This particular meeting was about using toxicology and new toxicology approaches . . . early on in the chemistry design process," she said. "To advance green chemistry we need to have a lot more interaction between the toxicologists and the chemists, not unlike what happens in the pharmaceutical industry."

Shelton-Davenport repeated several interesting comments that had been made at the workshop. "Richard Denison talked about how the new high-throughput, high-content data . . . should allow more assessment near the beginning of the chemical design process," although she noted that "high throughput" might be a bit of a misnomer because "high-throughput, high-content data isn't always rapid to analyze."

The idea underlying the meeting was that "chemicals can be designed to be inherently safer, which is the mantra of the green chemistry world," she said, and the meeting "was trying to get at what toxicologists know that could inform that." However, she added that it is just as important to understand the limitations of what toxicologists know.

"We had Thaddeus Schug from NIEHS talk about tiered endocrine disruption processes and testing," Shelton-Davenport continued. Robert Tanguay and Jim Hutchison spoke about the importance of using simple

---

[5] Further information on the Design and Evaluation of Safer Chemical Substitutions: A Framework to Inform Government and Industry Decisions is available at http://www8.nationalacademies.org/cp/projectview.aspx?key=49569 (accessed April 2, 2014).

organisms, such as zebrafish, to complement in vitro studies and tests in rodents. The two also described an interesting example of chemical design people working quite closely with toxicologists on nano-materials. Alex Tropsha discussed the importance of new approaches for combining short-term biologic assays to inform the structure–activity relationship; in particular, the goal was not just to look at structure and modeling but also to inform that with in vitro assays.

Then Shelton-Davenport switched to describing the study she is currently directing. The study is being funded by the EPA's Office of Research and Development, and it has its roots in the existence of many different approaches for comparing chemical substitutes. Shelton-Davenport listed just a few of these approaches: GreenScreen, Cleangredients, GreenList from SC Johnson, IC2 out of the Interstate Chemicals Clearinghouse, the EPA's Design for the Environment, California's Green Chemistry Initiative, Greenlist, Greenblue, Cradle to Cradle, SubsPort, and so on. "These are just a few of the different approaches," she said.

The various approaches differ in many ways. They differ, for instance, in how they consider health and safety effects versus ecological risks, such as aquatic toxicity or the environmental impacts of chemicals, when they are comparing alternatives. "To give an example," she said, "Cradle to Cradle is one that includes everything from environmental impact, as in greenhouse gases and water use, to social fairness. I think that is a pretty broad number of things to include. Some of the others are more focused on hazard or safety." They also differ in how they handle uncertainty and what they do when there are no data. And some of them are not very transparent so that it is difficult to know what goes into the alternative analysis.

In the new study that she is directing, the goal is to put together a more universally accepted approach to evaluating substitutions, Shelton-Davenport said. "I think most people would agree that if there was some harmonization of—or at least understanding about—the different kinds of approaches and the appropriateness of them for different uses, that would allow wider use of this comparison of chemical options." Furthermore, a more universally accepted approach should make it easier to plan for developing the scientific information and the tools that will be required for such an approach, and it should also help increase the dialogue among different stakeholders by having them all on the same page concerning which approach to use.

The committee's statement of task calls for it to develop a decision framework for the evaluation of potentially safer substitute chemicals. That framework should

- support the consideration of potential impacts early in chemical design;
- consider both human health and ecological risks;
- integrate multiple and diverse data streams;
- include details on how to consider trade-offs between risks and factors such as product functionality, product efficacy, process safety, and resource use; and
- identify the scientific information and tools required for the approach.

The committee is also charged with developing at least two examples that "demonstrate how the framework can be applied by different users in contrasting decision contexts with diverse priorities." According to the statement of task, these examples "shall include demonstration of how high-throughput and high-content data streams could inform assessment of potentially safer substitutes early in the chemical development process."

DISCUSSION

Lynn Goldman noted that there have been many recent scientific advances with implications for risk assessment and exposure assessment. Will these scientific advances translate into faster, more efficient assessments?

"I think that there is tremendous opportunity," Ila Cote said. Thanks to advances in personalized medicine and pharmacology, the field of risk assessment is moving forward quickly, she said. "My concern is that the toxicology community will be left behind. I think there is going to be progress that is going to be made whether the conventional community does anything or not."

Paul Gilman added, "I think the real challenge is in bringing along the different customers, everybody within the agency, the regulated community, and folks who want to be engaged." Understanding toxicology at the molecular systems level is a very rapidly moving field, he said. The researchers who are working in more predictive exposure analysis and who are feeding back and forth between structure and likely exposure scenarios "are talking in a language that has always been difficult to engage the community with." And now the community is also being asked to engage in informatics and computing and biology at the molecular level. "That is a real challenge, I think."

"I agree with that," John Balbus said. "The scientific advancements and technology sites tend to draw us toward the more involved and more complex. It is the political and social side that would be moving us toward a more streamlined process for a lot of decisions."

Jerry Paulsen of George Washington University suggested that it would be valuable to think beyond substitution when trying to minimize risk from chemicals. "Do we really always need alternative chemicals? Sometimes maybe we don't need chemicals at all. . . . Do we really need to spend a lot of time looking for alternative chemicals to be flame retardants in furniture when it is not clear that we need flame retardants in furniture? . . . Do we need fragrances for consumer products at all? Why do we spend the time looking for safer chemicals when the safest might be none at all?"

Balbus agreed that these are the sorts of questions that should be asked. "Do we need to be exposing people to this? What is the societal end we are trying to get, and what are the different ways to get there, and does it need to be something that involves chemical exposure?" These questions should be part of the decision-making framework, he said.

Marilee Shelton-Davenport said she believes this issue will come up in the committee study she is directing. "The title is Evaluation of Safer Chemical Substitutions, but my thinking is that they are likely going to be having a broader discussion."

Gina Solomon of the California Environmental Protection Agency also weighed in on the issue. "In the California Safer Consumer Products Regulations, we have devised an off-ramp where if the state identifies a chemical of concern in a product and lists it, the manufacturer of that product may either perform an alternatives analysis or may simply take the chemical out of the product, and then that would save them the trouble of having to go through the entire alternatives analysis. We will see what comes of that and whether that does incentivize removal of some of these chemicals altogether."

REFERENCES

EPA (U.S. Environmental Protection Agency). 2009. *The U.S. Environmental Protection Agency's strategic plan for evaluating the toxicity of chemicals.* Washington, DC: Office of the Science Advisor, Science Policy Council. Available at http://www.epa.gov/spc/toxicitytesting/docs/toxtest_strategy _032309.pdf (accessed March 31, 2014).

EPA. 2011a. *Advancing the next generation (NexGen) of risk assessment: public dialogue conference.* Washington, DC: EPA. Available at http://www.epa.gov/risk/nexgen/docs/NexGen-Public-Conf-Summary.pdf (accessed March 31, 2014).

EPA. 2011b. *Advancing the next generation (NexGen) of risk assessment: The prototypes workshop.* Washington, DC: EPA. Available at http://www.epa. gov/risk/nexgen/docs/NexGen-Prototypes-Workshop-Summary.pdf (accessed March 31, 2014).

EPA. 2013. *Next generation risk assessment: Incorporation of recent advances in molecular, computational, and systems biology (external review draft).* Available at http://cfpub.epa.gov/ncea/risk/recordisplay.cfm?deid=259936 (accessed April 2, 2014).

NRC (National Research Council). 1983. *Risk assessment in the federal government: Managing the process.* Washington, DC: National Academy Press.

NRC. 2007. *Toxicity testing in the 21st century: A vision and a strategy.* Washington, DC: The National Academies Press.

NRC. 2009. *Science and decisions: Advancing risk assessment.* Washington, DC: The National Academies Press.

NRC. 2012. *Exposure science in the 21st century: A vision and a strategy.* Washington, DC: The National Academies Press.

5

Approaches to Prioritizing Chemicals for Risk Assessment and Risk Management

Roundtable member and session moderator Andrew Maguire opened the fourth session and suggested that the overall theme of the session was innovation. "We see a bewildering array of levels and types of what we know or think we know and also what we don't know," he said. "We see that there are very long cycles for decisions. Is the data relevant enough or not? Is it sufficiently conclusive or not for policy making? . . . For this panel, the question is, What can we do now? What can we do soon enough to protect the health of many millions of people? How do we make the decisions that we need to make? How do we move ahead?"

In particular, the topic of the session was examples of improved approaches to priority setting in the risk assessment and risk management of chemicals. A good name for the panel, Maguire suggested, would be the Who's Doing What Panel, as each panel member was to describe what his or her institution is doing to improve the prioritization of chemicals for testing and management. The four institutions represented on the panel were the U.S. Environmental Protection Agency (EPA), the California Environmental Protection Agency (California EPA), Health Canada, and the American Chemistry Council (ACC).

DEVELOPING MODELS TO PRIORITIZE CHEMICALS FOR TARGET TESTING

The session's first presenter was Richard Judson of the National Center for Computational Toxicology at the EPA. He spoke about the use of toxicological models to prioritize chemicals for testing.

73

Judson began by noting that although he had been asked to speak about ToxCast[TM],[1] the EPA's program for forecasting toxicity, his talk would actually cover a much larger collaborative effort among the EPA and a number of other groups. "We have collaborators at the National Toxicology Program and elsewhere at the NIH [National Institutes of Health], the FDA [Food and Drug Administration], and a number of academic and government and industry collaborators, both here and in Europe," he said. "It is a very big collaboration where we are all working on different parts of this problem of solving the too many chemicals problem. It is not just us."

The EPA's computational toxicology center was founded nine years ago to deal with the "too many chemicals" problem. "We know that there are lots of chemicals out there," he said. "Most of them have never been tested in the standard animal tests. A critical issue is: How can we take thousands of chemicals and run a consistent set of tests on them?" A second issue is how to use in vitro methods combined with the results from tests on laboratory animals to infer the human risks of those chemicals.

"We and our collaborators have come up with an overall strategy," he said. The basic idea is to develop a better understanding of the modes of action of these chemicals in the human body and thus to be able to use data from tests on laboratory animals and elsewhere to build models that can be used to predict computationally the effects of various chemicals on humans. There are four basic steps in this strategy: to develop high-throughput in vitro assays for testing chemicals on the biological pathways linked to toxicity, to use that information to develop predictive hazard models, to develop high-throughput exposure predictions, and to create the data and models to assess risk from the chemicals. These models can be used to prioritize chemicals for targeted testing, to distinguish possible adverse outcome pathways for chemicals, and to provide semiquantitative high-throughput risk assessments.

Judson then went into detail on each of these steps in the computational toxicology approach. Before beginning, it is necessary to identify the various biochemical pathways through which chemicals exert their influence in the human body. Generally speaking, the chemicals work by perturbing the usual function of a pathway in some way that leads to a toxic effect. Thanks to a great deal of work by scientists in a variety of fields, many of these pathways have already been identified.

---

[1] Further information on ToxCast[TM] is available at http://www.epa.gov/ncct/toxcast (accessed March 31, 2014).

With this information in hand, the next step in the program is to develop high-throughput in vitro assays to test for the effects of various chemicals on these biochemical pathways. The effects of the chemicals can be tested on proteins or in whole cells or even in something as complicated as zebrafish. "You can actually treat whole zebrafish in a microtiter plate well," Judson said. He noted that researchers can detect a variety of phenotypes using this approach. In particular, zebrafish provide a good model for detecting chemicals causing a range of developmental defects. Currently, several laboratories have tested thousands of chemicals in these sorts of assays.

The EPA did not have to pay for most of the technology development for the assays, Judson said, because most of the development had already been done by the pharmaceutical industry. "They spend a billion dollars developing every drug," he said. "They have made investments in basic technology, which we were then able to buy off the shelf."

There are many different types of assays that can be run, Judson said. "You can use models—you just start with structure and run it on the computer. There are assays that are run in cells, assays run in zebrafish, or in *C. elegans*." At one extreme, he explained, we can run cell-based assays that tell us what is going on with every one of the 20,000 genes in the whole genome.

At the moment, he said, these whole-genome assays are too expensive—thousands of dollars per chemical—to do for every chemical. However, "The Broad Institute, 2 weeks from now, is releasing its first big dataset with a version of this chip that is getting down to hundreds of dollars per chemical. That puts doing whole-genome analysis for every chemical that we are exposed to within the realm of possibility."

Running hundreds of assays on thousands of chemicals results in huge datasets. The largest dataset that has been generated in the cross-agency Tox21[2] collaboration has about 8,200 chemicals in it, including both environmental chemicals and chemicals found in consumer products (Zang et al., 2013). There are a number of flame retardants and also drugs, food additives, chemicals in consumer products that are put on the skin, and so forth.

With those data in hand, the next step is to create models that can accurately predict what happens when people are exposed to the chemicals.

---

[2] Further information on Tox21 is available at http://epa.gov/ncct/Tox21 (accessed March 31, 2014).

The most basic approach is to take in vitro data and in vivo data and apply statistical analyses to look for relationships between the two sets of data; ideally, the relationships would make it possible to use in vitro data to predict the effects on humans of chemicals for which there are no in vivo data. "In the most basic approach, you treat all of this in vitro data as just a bunch of numbers," Judson explained. "The chemical either turns on that pathway or not. Likewise, you treat all of the in vivo data as just another bunch of numbers. This chemical either causes cancer or it doesn't. It is very easy to do. A lot of groups have done this." Unfortunately, Judson continued, this approach does not work particularly well. Even with all the data that are available, the data do not provide enough statistical power to reliably identify relationships between the chemicals and their effects in living organisms.

"What you have to do is actually put some biological knowledge into the mix," he said. "We just don't have enough data to let statistics drive everything."

One way of inserting biology into the models is called the adverse outcome pathway approach. This involves searching the scientific literature for biochemical pathways involved in the adverse outcomes created by a particular chemical. To describe the approach, Judson offered a hypothetical example involving vascular disruptor chemicals, or VDCs (see Figure 5-1). "We gave this problem to a really good postdoc, telling her, 'We have some evidence that if you perturb VEGF [which is a particular pathway that is involved in vasculogenesis in a developing embryo], you can cause certain kinds of developmental defects.' Our smart postdoc then works out, by going to the literature, that VEGF actually disrupts cytoskeletal signaling in endothelial cells. If the chemical does that, then it leads to specific whole-animal defects." By scanning the literature one can come up with multiple pathways leading from an initial chemical interaction with a pathway to something happening in cells to something happening in tissues to something happening in the whole organism.

Environmental chemicals cause toxicity by interacting with specific biological molecules, Judson noted, so one can examine the various biological molecules involved in a particular adverse outcome pathway—for example, VEGF or CCL2 or the aryl hydrocarbon receptor in the case of embryonic vascular disruption—and examine the in vitro databases to see which environmental chemicals affect those biological molecules. "You can now start doing a reasonable job of predicting the kinds of chemicals that we have tested that might have these kinds of phenotypes."

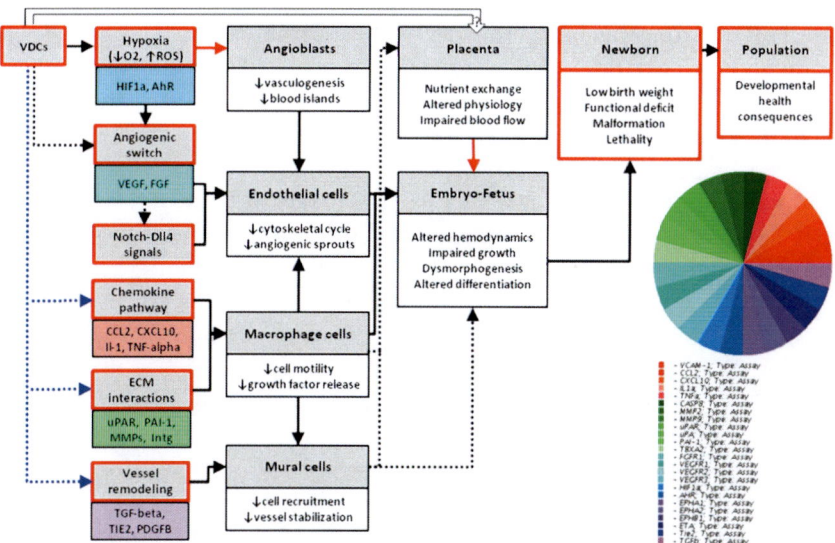

FIGURE 5-1 Adverse outcome pathway approach.
SOURCE: Knudsen and Kleinstreuer, 2011. Reprinted with permission from John Wiley & Sons, Inc.

There are now some very detailed and complicated experiments called "targeted testing experiments" being done in collaboration with a number of academic and industry groups in an effort to work out specific pathways for specific chemicals, Judson said. "This has shown that you don't have to solve all the problems for all chemicals, but by having this big database, you can take specific chemicals and understand a lot of detail of what is going on all the way to the toxic endpoint."

It is also possible to model the effects of chemicals on groups of cells using what is called a "virtual tissue model." With such a model one can simulate the behavior of a group of cells or a bit of tissue over time—say, the tissue in a developing limb bud. The model takes into account the various biochemical pathways in the cells and between the cells that control the development of the limb bud, and so it becomes possible to observe how a particular environmental chemical affects that development. "You can actually look at the dynamics at the level of cells and groups of cells," Judson said.

By combining biological information with in vitro and in vivo data in this way, it opens up a new approach to prioritization, Judson said. "Once you have those models, then you can take a new chemical where you don't have animal data and you can run simple assays and make a prediction. Does this look like it might be a carcinogen? Does it look like it might be a developmental toxicant?" The predictions are not 100 percent accurate, he said, but they are accurate enough to do prioritization. "We are not trying to replace animal tests yet. Maybe one day. But if we want to do prioritization, this approach looks good enough."

In addition to knowing which biochemical pathways are affected by which environmental chemicals, it is also important to know what dose of a particular chemical is necessary to produce an effect. The EPA has developed an approach to creating estimates for what it calls the biological pathway altering dose, which is the amount of a chemical necessary to turn on a particular pathway.

"This approach requires two experiments," Judson explained. "You measure the intrinsic clearance rate of a chemical in liver cells, which could either be a rodent or human, and you measure plasma protein binding. These parameters are then used in a relatively simple computer model that calculates what we call the Css, the Concentration at Steady State per daily dose, which is just the conversion factor: If I take 1 milligram per kilogram per day of a chemical, what is my steady state concentration going to be?" It is also possible to go beyond steady state assumptions. Finally, he said, the in vitro potency is combined with the Css values, and this produces the biological pathway altering dose.

This process is all done in vitro, and it costs perhaps $1,000 to get a reference dose for a particular pathway. However, Judson offered an important caveat: "It is still just a concept. We have shown a couple of cases where you get within about an order of magnitude of what you would get from animal studies." But it is not yet a fully proven approach.

Once there is an understanding of the biological effects of chemicals and the dosage at which the effects take place, it is necessary to get information about exposure, Judson said. "Hazard doesn't mean anything if you don't have some estimate of exposure," he said. "You need to have exposure models that are equally high throughput. You have to be able to make some sort of prediction for thousands of chemicals."

To that end the EPA created the ExpoCast[3] program for high-throughput modeling of exposure. It is well behind the efforts to model toxicity, he said, but the agency has shown it is possible to make exposure estimates for 10,000 compounds. "Those exposure estimates

---

[3] Further information about ExpoCast is available at http://www.epa.gov/ncct/expocast (accessed March 31, 2014).

have really wide confidence intervals today," Judson said. The agency will be making a big effort over the next few years to reduce those confidence intervals, he added, but even with today's confidence intervals it is possible to use the estimates for prioritizing chemicals for further testing.

One of the most important parts of understanding exposure is knowing how a chemical is used, Judson noted. This fact led the EPA to develop the Chemical and Product Categories (CPCat) database, which contains information on about 40,000 compounds.[4]

Summing up the work on ToxCast, he said that the program has been controversial with some audiences, with some recent presentations and publications claiming that ToxCast has failed. "There are problems," he acknowledged. "Having said that, though, the EPA is confident that you can use this approach for prioritization. We are not going to ban a chemical or say that a chemical is really bad or really good because of this, but we can start prioritizing chemicals for further testing."

The first real-world application of ToxCast is likely to be the Endocrine Disruptor Screening Program, a congressionally mandated program to put about 5,000 chemicals through a screening process. The problem is that the screen costs about $1 million per chemical, and the worldwide testing capacity is only about 50 to 100 chemicals per year. Thus, with current technology, it would cost $5 billion and take 50 to 100 years to complete.

The ToxCast approach should make the screening program feasible. The relevant pathways are known, and the exposure assessment can be done. The EPA has already started, Judson said, and the first outcomes should be available in about 3 years.

APPROACHES TO PRIORITY SETTING IN CALIFORNIA

Gina Solomon, Deputy Secretary for Science and Health at the California EPA, spoke about how the California EPA is setting priorities on environmental chemicals. The state is not trying to compete with federal programs, such as those at the EPA, she said. Instead her agency is trying to develop complementary programs that can "help move

---

[4] Further information on the CPCat database is available at http://actor.epa.gov/cpcat/faces/basicInfo.xhtml (accessed March 31, 2014).

forward issues around prioritizing chemicals, identifying issues of concern, and taking action when necessary."

There are different approaches to priority setting depending on what one is actually doing, Solomon said—for instance, whether one is interested in doing screening or doing assessments. To illustrate the sorts of priority setting that are taking place in California, she described three statewide programs: CalEnviroScreen, Biomonitoring California, and Safer Consumer Product Regulations, formerly known as Green Chemistry. Each program sets priorities in its own way.

CalEnviroScreen Program

CalEnviroScreen[5] is an environmental justice screening tool, created by the Office of Environmental Health Hazard Assessment, that is used statewide (California EPA and OEHHA, 2013). Version 1.0 was released in spring 2013, and version 2.0 was released in August 2014. Solomon explained that the screening tool contains 17 different indicators, including such things as pesticide use; various air quality indicators; indicators of environmental effects, such as leaking underground storage tanks and toxic cleanup sites; and vulnerability factors, which range from poverty, educational attainment, and linguistic isolation to asthma emergency room visits and low birth weight. The data are mapped to individual zip codes, and they will soon be mapped to individual census tracts.

Total pollution scores are calculated from the indicators, with the environmental effects indicators being multiplied by a factor of one-half compared to the human exposure indicators, and the pollution scores are then placed into deciles, said Solomon. These pollution scores are next multiplied by the population vulnerabilities, with the result being a risk score for each zip code in the state. Figure 5-2 is a map of California with the areas in the top 5 percent of risk indicated in blue and the next 5 percent in orange. As can be seen from the figure, many of the highest risk areas of the state lie in the Central Valley, the Imperial Valley, and around Los Angeles.

---

[5] CalEnviroScreen is available at http://oehha.maps.arcgis.com/apps/OnePane/basicviewer/index.html?appid=1d202d7d9dc84120ba5aac97f8b39c56 (accessed April 4, 2014).

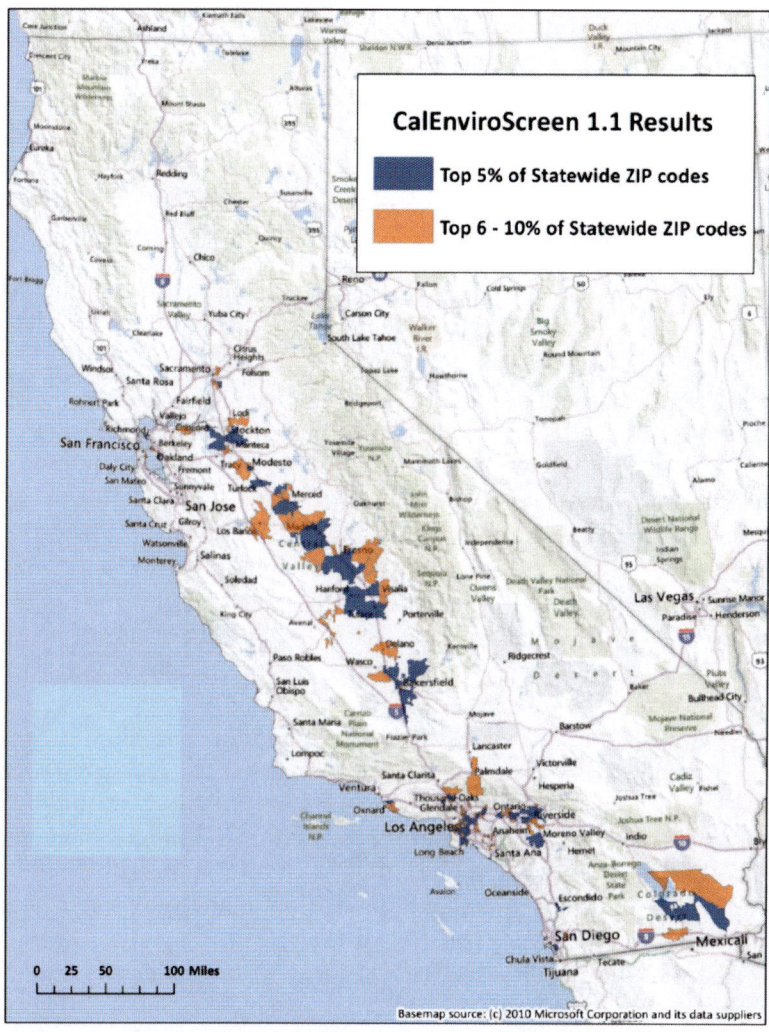

FIGURE 5-2 CalEnviroScreen 1.1 results: Mapping application of zip codes with highest CalEnviroScreen 1.1 scores.
SOURCE: California OEHHA, 2014. Reprinted with permission from California OEHHA.

What does this sort of environmental justice screening tool have to say about chemical priority setting? Solomon said that using such a tool to focus assessments on disproportionately impacted communities can be important in pointing to chemicals that deserve further scrutiny. For example, she said, perchlorate was not identified as an important chemical until it was found in the drinking water supply of the town of Rancho Cordova in California in 1997 after the Department of Public Health developed a new test that had a lower limit of detection. "This was a community that was already known to be impacted from a Superfund site," she said. "Testing the water supply was a reasonable thing to do in honing down to look at the risks to that community. Lo and behold, this chemical has now become a fairly significant priority nationwide."

Biomonitoring California

The second program that Solomon described was the Biomonitoring California program,[6] which was established in large part in response to the detection of polybrominated diphenyl ether (PBDE) flame retardants in breast milk. These chemicals were first identified in Sweden in the late 1990s, but soon after the Swedish studies were published, the California Department of Toxic Substances Control laboratory adopted a method for monitoring for the chemicals and reported detecting them in two settings. They were found in harbor seals that had died and washed up in the San Francisco Bay and also in breast biopsy specimens from women in the San Francisco Bay area.

The most striking and worrisome part of the finding was that the levels measured in the human samples from around the San Francisco Bay were 40 times higher than the levels from the Swedish study. "This got attention in the media and in the legislature," Solomon said. "Some of the PBDEs were banned the following year in California."

The push to establish the California Biomonitoring program was the result of the "realization that biomonitoring can be useful for identifying new priorities, new chemicals that we weren't really thinking about," she said.

There are two aspects to the chemical selection and priority setting in Biomonitoring California. The program starts with the chemicals that the Centers for Disease Control and Prevention (CDC) is looking at. But it also has a scientific guidance panel that can designate and prioritize chemicals that are outside the CDC biomonitoring program list. "We have tended to focus on chemicals that are outside the list," Solomon

---

[6] Further information on the Biomonitoring California program is available at http://www.biomonitoring.ca.gov (accessed April 2, 2014).

said, "because we feel that the CDC is doing a very excellent job on what they are working on. We want to be value added."

Thus, one of the things that the scientific guidance panel has prioritized is chemicals whose level of use may be different in California than in other states and, particularly, those chemicals whose use may be increasing in California. Thus, the panel has examined chemicals that are serving as replacements for chemicals that have been banned or restricted in California. Among the categories of chemicals that are shifting in California are flame retardants, phthalates and bisphenol A in several uses, and perchloroethylene in dry cleaning, whose use is being phased out across the state.

"That has resulted in a number of chemicals—and, more specifically, chemical groups—being listed as priorities for the California Biomonitoring program," Solomon said. Thus, the scientific guidance panel designated several categories of flame retardants as a group even though they are somewhat different chemicals structurally, because they are used in similar niches. "There is going to have to be additional prioritization within those," she added, "but our laboratories are working on developing methods to screen for as many of those as they can."

The panel is also looking at a broad collection of various bisphenols, some of which are potential replacements for bisphenol A and others of which have somewhat different uses. The various chemicals were grouped together into a chemical class, Solomon said.

There are some categories of chemicals that the panel decided not to prioritize, she said, such as synthetic hormones used in food production and antimicrobials used in food production. The reasons for not prioritizing the various categories included because the chemicals in a category were too disparate, because the chemicals were too difficult to biomonitor effectively, or because there was no clear California-specific concern related to the chemicals.

Most recently, Biomonitoring California is preparing to do non-targeted testing with a time-of-flight mass spectrophotometer, which allows screening for unknown chemicals—those that are not specifically included in current biomonitoring assays—in environmental or biological samples. Nontargeted screening can help identify potential exposures and set priorities for testing, risk assessment, or ultimately mitigation. "If you sample a whole lot of people and start consistently seeing something that you weren't expecting to see," Solomon said, "it is time to develop a method and start specifically looking for that in populations of interest, evaluating the risk, and deciding if further action is needed. It is definitely a very useful tool for priority setting."

California Safer Consumer Product Regulations

The third program is California Safer Consumer Product Regulations[7] under the Department of Toxic Substances Control. The initial chemical priority setting in that program largely relies on lists from others. The program currently uses 23 lists containing approximately 1,200 chemicals or chemical groups, she said. "This is more of a risk management kind of program," she explained. "We are not trying to reinvent the chemical priority-setting wheel, here. We are trying to focus on prioritizing at the product level, getting companies to look at alternatives and ultimately to move gently towards safer alternatives."

From this initial list of about 1,200 candidate chemicals, the program is focusing on about 200 chemicals for the initial round of product selection, and based on these 200 it will choose a set of up to five priority consumer products. For each of these priority products there must be potential exposure to the candidate chemicals and also the potential for the exposures to contribute to or to cause significant or widespread adverse impacts.

Looking for a chemical in a product is especially tricky, Solomon said, because there is so little information available on which chemicals are being used where. "I have lots of examples of people in our agency who are phoning up companies saying, 'I want to get a quote on this. Can you tell me what products you use? Can you send me the names of what you are actually selling?' Or else they are going to stores and looking at what is on the shelf. We are just scrambling to figure out what is used in California in what amounts."

The initial list of five priority products will be just the beginning, Solomon said. After that list is released, the program will be putting out a work plan and developing a much more ambitious list.

ASSESSING AND PRIORITIZING RISKS IN CANADA

The next presenter was Heather Patterson, Senior Evaluator in the Healthy Environments and Consumer Safety Branch of Health Canada,

---

[7] Further information on California Safer Consumer Product Regulations is available at http://www.dtsc.ca.gov/SCP (accessed April 2, 2014).

which is Canada's equivalent of the U.S. Department of Health and Human Services. She is currently working on developing innovative approaches for prioritization and assessment, and she described efforts in Canada to assess and prioritize chemicals.

The legislation underpinning Canada's efforts in that area, Patterson explained, is the Canadian Environmental Protection Act, which was passed in 1988 and amended in 1999. It provides the regulatory framework in Canada for information collection, risk assessment, and risk management of new and existing chemicals and organisms. It includes provisions for the assessment of existing chemicals, she said, and it requires that every new substance made in Canada or imported into Canada be assessed against specific criteria. According to the act, the Minister of the Environment must maintain an inventory of existing substances in Canada, known as the Domestic Substances List (DSL), and this list is the sole basis for determining whether a substance is deemed to be new or existing in Canada.

The 1999 amendments required the ministers of the environment and health to categorize the approximately 23,000 substances that were on the DSL, using specific criteria to identify priorities for future assessment work. Patterson noted that Environment Canada looked to see which of those 23,000 substances had the potential to be either persistent or bioaccumulative and which of those were inherently toxic to non-human organisms, while Health Canada looked to see which of the substances posed the greatest potential for human exposure and which ones were likely inherently toxic to humans.

In Health Canada's work determining which substances had the greatest potential for human exposure, it focused on three lines of evidence, Patterson said. It looked at the amount of a substance in commerce, the number of identified companies manufacturing or importing a given substance, and the use codes or the uses of the substances that were identified. "This was based on the data that was given to us from 1984 to 1986," she added. "It was old even at the time of categorization, but it was the only information that we had for every substance on the DSL so we could compare equally across all of them."

To identify which substances for further assessment were potentially toxic to humans, Patterson said, the agency performed a "list-matching exercise," looking to see which substances had been classified by other agencies as carcinogens, mutagens, or reproductive toxicants. When Health Canada combined its priorities with those identified by Environment

Canada, the results identified 4,300 substances for future assessment work.

As this categorization work was being wrapped up, Canada's Chemicals Management Plan was announced.[8] That plan, which is the government of Canada's response to the *Strategic Approach to International Chemicals Management*, is designed to meet the 2020 goals set by the World Summit on Sustainable Development for the sound management of chemicals. She noted that it provides a framework for assessment and for the management of the priorities identified through categorization, and it integrates multiple federal programs into a single strategy to ensure that the chemicals are managed appropriately in order to prevent harm to Canadians and their environment.

The Chemicals Management Plan has three phases: 2006–2011, 2011–2016, and 2016–2020. In the first phase, she said, the highest-priority substances were addressed and work was initiated on the lower-priority substances. The major focus of the second phase, which is now ongoing, is the substance groupings initiative. In the third phase the remainder of the 4,300 priorities will be addressed.

Patterson explained that the highest-priority substances dealt with in the first phase of the plan were about 500 substances that met the criteria of potential persistence, bioaccumulation, and inherent toxicity to aquatic organisms that had been identified by Environment Canada, or that had high exposure potential, and that were identified as posing a high hazard for human health. These substances were addressed through three different mechanisms.

There were about 150 substances that were persistent, bioaccumulative, and inherently toxic but that were believed to no longer be in commerce in Canada, Patterson said. "For those substances, we published SNAcs. This is a Significant New Activity. It is very similar to a SNUR [significant new use rule] in the U.S., which means that if anyone wants to use these substances for a new activity, they need to notify us before they can do so."

There were about 200 substances that were of high concern and were found to be in commerce, she said. These were assessed under the Challenge Program, which consisted of a variety of individual screening-level assessments of the substances.

---

[8] Further information on the Chemicals Management Plan is available at http://www.chemicalsubstanceschimiques.gc.ca/plan/index-eng.php (accessed April 2, 2014).

Finally, there were about 160 substances of high concern that were found to be used predominantly within the petroleum sector. A focused sectorial approach was developed to deal with these substances that relied on exposure-based prioritization (see Figure 5-3).

Depending on a substance's characteristics—such as whether the substance left the facility where it was produced and whether it was used by the public, only by industry, or by other sectors—the substance was classified as being in Stream 0, Stream 1, Stream 2, Stream 3, or Stream 4 (Stream 0 was used for products that were similar in composition to petroleum substances, but were not manufactured or used by the petroleum sector). "The assessment approaches that we used to deal with these substances were tailored according to the exposure scenarios identified for each stream," Patterson said.

In addition to its work on the high-priority substances, Health Canada was also identifying low-priority substances. This is a different sort of exposure-based prioritization method, Patterson noted. In identifying low-priority substances, the premise was "no exposure equals no risk," she said. "We wanted to focus on this early in the Chemicals Management Plan because it allows us to identify substances with low concern so that we can focus our resources on those substances with higher concern." It turned out that a number of substances that had been identified through categorization as priorities actually had a low potential for exposure because

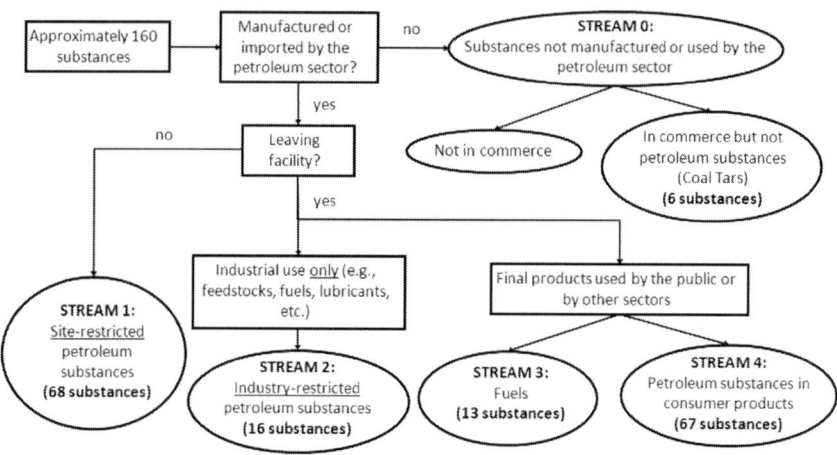

FIGURE 5-3 Petroleum-sector stream approach.
SOURCE: Patterson, 2013.

very low volumes of the substances were manufactured or used in Canada. "Environment Canada addressed these substances by doing worst-case modeling to calculate predicted environmental concentrations and compared those with toxicity values," she said. "At Health Canada, we did not quantify exposure, but rather we identified which substances had no or very low exposure potential."

Exposure could be direct or indirect. By their nature, those substances that were used in very low volumes had low potential for indirect exposure via the environment. Thus, the main issue was the potential for direct exposure. In particular, was a substance used in consumer products? "For substances that are used in consumer products," she said, "even if there is only a low volume in commerce, there could still be a high potential for exposure if the substance is applied directly to your skin, for example."

For those substances found to not have the potential for indirect exposure and to not be used in consumer products, Health Canada concluded that further assessment work was not necessary. After carrying out three assessments on a total of approximately 1,200 substances, which were published in 2013, the agency found that approximately 700 of the substances needed no further assessment work. The other 500 will be considered further in future assessments, Patterson said. "We expect that we will be using this approach again to deal with as many substances as possible once we collect current commercial status on the remaining priorities."

As the first phase of the Chemicals Management Plan was being wrapped up, those who were responsible for its implementation began working on priorities for the second phase, Patterson said, and what became obvious was that doing assessments of single individual substances was not always the most efficient approach. So they developed an approach by which they would collect similar substances into groups and use those groupings in their prioritization and assessment work.

The groupings were developed based on a number of different factors: common chemical classes, common modes of action, common uses, common sectors, and so on. "We also considered scheduling implications," she said. "We wanted to look for the availability and timing of international information. For example, if we are expecting a major assessment report from the United States, we would like to time our assessment report to either be closely related with that or maybe even follow it to allow for consideration of all possible information." They also considered the ability of various stakeholders to participate. "We tried not to align all of our metals groupings, for example, at the exact same time because that sector would be overwhelmed with having to provide data and comments."

The result was the Substance Groupings Initiative, which covers about 500 substances divided into nine groups, such as aromatic azo- and benzidine-based substances (which is by far the biggest grouping, with more than 300 substances), cobalt-containing substances, certain organic flame retardants, and phthalates. One grouping, labeled Certain Internationally Classified Substances, is not actually a group of similar substances; rather it is a group of individual substances that the planners felt warranted attention due to their international high hazard classifications.

Pulling back and looking at all 4,300 substances requiring assessment as a result of categorization, Patterson offered an overview of the different assessment approaches being used (see Figure 5-4). As described above, there were 500 highest-priority substances that were divided into SNAcs, Challenge Substances, and petroleum-sector substances; 700 that needed no further assessment because there was low potential for exposure to them; and 500 in the Substances Groupings Initiative. She says there are likely to be another 200 or so substances that will fit into the petroleum-sector stream approach and another 700 substances or so that will be found to have low exposure potential after further data collection. "We also have about 700 polymer substances that we are dealing with at this time," she said. "We are developing an approach to deal specifically with those substances." That leaves about 1,000 more substances for which the assessment approach has not yet been determined.

Now that Health Canada is moving toward the middle of the second phase of the plan, it is time to think about how to approach the third phase, Patterson said. "It is not so much about selection anymore," she said. "We have to figure out what the best way is to deal with what is left." To do that she and her colleagues are again looking at how to group these substances based on such things as structure, mode of action, functional use, and possible substitutions. They are also looking at what the potential exposures are for each of the substances. One important issue is the commercial status in Canada for each substance. How much of it is in commerce, and how is it being used? Is it used in consumer products? And has it been observed in environmental media or biomonitoring studies?

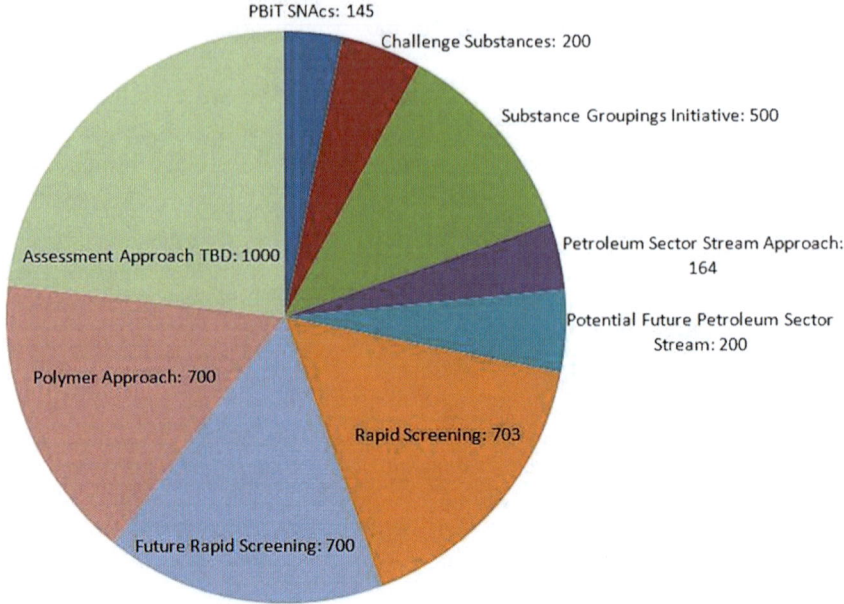

PBiT SNAcs: 145
Challenge Substances: 200
Substance Groupings Initiative: 500
Petroleum Sector Stream Approach: 164
Potential Future Petroleum Sector Stream: 200
Assessment Approach TBD: 1000
Polymer Approach: 700
Rapid Screening: 703
Future Rapid Screening: 700

FIGURE 5-4 Assessment approaches for categorization priorities.
NOTE: PBiT = persistent, bioaccumulative, and inherently toxic, SNAcs = significant new activities, TBD = to be determined.
SOURCE: Patterson, 2013.

Another issue is the availability of toxicity data. Is there empirical data on a given substance? How much? Is it positive or is it negative? What end points are involved? Is there more than a median lethal dose (LD_{50})[9] available? Are there high-throughput screening data available? As an example, she described the results of a TOXLINE search for data on the 2,300 remaining priority substances. Fifty-three percent of the substances had no hits at all, and 18 percent had very few hits, and since many of the hits for those 18 percent are likely to not be relevant to human health, there may be only 29 percent of the 2,300 remaining priority substances that have empirical data relevant to the assessment of human health. "I know this is only one source," she said. "There may be

---

[9] Median lethal dose (LD_{50}) is the statistically derived median dose of a chemical or physical agent expected to kill 50 percent of organisms in a given population under a defined set of conditions (IUPAC, 2007).

other data out there and we are still gathering data. But this really highlights the need for innovative assessment approaches."

Finally, there are scheduling issues. These include the identification of possible opportunities for international collaboration and the alignment with data generation. For example, she said, Richard Judson is "coming up with some great new high-throughput ideas." She noted that we need to wait until we get those data and determine how best to use them before beginning to assess substances for which that is the only source of toxicological data.

In carrying out assessments, it is important to identify emerging priorities, Patterson said. It is well recognized, for instance, that the scientific understanding of exposure and toxicity continues to evolve over time and that global regulatory action on chemicals also changes over time. "So we can't just stop with using the categorization decisions to decide on priorities," she said. "We have to continue to update our list of priorities based on the evolving landscape."

Traditionally, she continued, there are seven "feeders" that are used for the identification of priorities: categorization decisions, industry information, decisions from other jurisdictions in Canada, international assessments or data collection, public nominations, trends in new substance notifications, and emerging science or monitoring data. "We are currently looking for a way to make this process more systematic so that we can ensure we are looking at all of the appropriate pieces of information in a timely manner," she said.

To conclude, Patterson offered a list of lessons learned to date:

- There are many limitations to conducting a priority-setting exercise that is based on dated inventory data, but it is often difficult to get new data.
- There is a lack of approaches available for modeling substances other than the generic organics. "When it comes to the inorganics, UVCBs [Unknown or Variable compositions, Complex reaction products and Biological materials], and polymers, for example, if we don't have empirical data, there is not a lot we can do."
- Indirect exposures, such as through environmental media, do not typically drive human health assessment outcomes.
- Instead, direct exposures—i.e., consumer product exposures— are more typically the key drivers in assessment outcomes. "We typically have to use upper-bounding models to develop scenarios for these substances. We refine them if the data are

available. What we have found is that it is often not easy to obtain data to refine these scenarios."

- A substance-by-substance approach is less efficient for both risk assessment and risk management.
- However, assessment of substance groupings can also be quite challenging. Groups built for one purpose are not always well suited for others. "If we build a grouping based on informed substitution, the risk assessment is often quite challenging because those substances could have very different exposure patterns or health implications."

In the discussion period that followed all of the session's presentations, Liz Harriman with the Massachusetts Toxics Use Reduction Institute asked Patterson about the benefits of grouping as well as any cautions. She answered that there are pros and cons to grouping. "Obviously, for the substances that have absolutely no data, building a group can allow read across from a data-rich substance to the data-poor substances within that group," she said. "That is how we are dealing with a lot of substances right now that have no empirical toxicity data available on them." On the other hand, she said, building a group for one purpose often makes a grouping quite detrimental for other purposes. "Building a group based on a common mode of action may make it almost impossible to do risk management in the end or vice versa," she said, "while building a group based on a sector makes it really difficult to do a risk assessment. Substances used in the same sector can still be used in very different ways."

AMERICAN CHEMISTRY COUNCIL VIEWS
ON CHEMICAL PRIORITIZATION

The final speaker was Christina Franz, Senior Director of Regulatory and Technical Affairs at the ACC. She described the ACC's views on chemical prioritization processes and on how to improve the prioritization process used by the EPA.

She began with some background on how the ACC came to develop a prioritization tool and the purpose of that tool. In 2009 the ACC published 10 principles for modernizing the Toxic Substances Control Act (TSCA) of 1976.[10] One of those principles called for the EPA to systematically

---

[10] Toxic Substances Control Act of 1976, Public Law 94-469, 94th Congress.

prioritize chemicals for safe use determinations. Coincidentally, during that same year the EPA also published Principles for TSCA Modernization, which called for manufacturers and the EPA to assess and act on priority chemicals.

Since that time Congress has held a number of hearings on bills proposed to modernize TSCA, and, Franz said, there has been a general consensus among the witnesses at these hearings that prioritization is essential. In 2012, the EPA came out with a list of 83 priority "work plan" substances that it announced it would be doing targeted risk assess-ments on from 2012 through 2015. And the Chemical Safety Improvement Act of 2013[11] became the first bill introduced in Congress to modernize TSCA that has included a prioritization section within the statute.

The ACC agrees with this emerging consensus on prioritization, Franz said. "It is clear that, from a practical standpoint, we have limited resources and limited time, and this requires that we focus on those substances that are of highest priority for further evaluation," she said. Thus, in particular, the ACC sees the EPA's work plan for a chemical prioritization process as an extremely important step forward for the agency. TSCA does not specifically direct the EPA to undertake such a prioritization, but it was within the agency's authority to do so.

When the EPA first published its proposed priority-setting approach, that approach focused exclusively on hazard, Franz said, and it ignored the exposure part of the prioritization equation. The ACC was pleased that, after listening to various stakeholders, the EPA decided to integrate hazard and exposure factors to identify substances for further evaluation. Still, she said, the ACC believes that the EPA's process can be improved—not for the 83 substances that have already been identified for targeted risk assessments, but rather in moving beyond those to a broader set of priorities for further evaluation.

"The process that the agency did employ was not sufficiently based on objective, science-based criteria that could be applied consistently across all chemicals evaluated," Franz said. "They began their prioritization assessment by looking at lists of chemicals that currently existed and then whether substances within those lists met certain factors that the agency was concerned about. The result, in ACC's view, is that there are inherent biases that exist. The priorities identified might not actually represent the highest hazard and greatest exposure potential and therefore could be a waste of time and resources on the part of the Agency."

---

[11] Chemical Safety Improvement Act of 2013, S. 1009, 113th Congress, 1st session (May 22, 2013).

In short, she said, prioritization should integrate hazard and exposure criteria. "If you have high exposure with no hazard, it shouldn't be a concern. Similarly, high hazard with no exposure, also not a concern. The highest priorities really exist at the intersection of highest hazard and greatest exposure."

In 2009 the ACC developed a prioritization tool that would embody these principles,[12] said Franz. It refined the tool in 2011 so that the tool could deal with chemicals lacking sufficient information for priority setting, would be better aligned with the Globally Harmonized System of Classification and Labeling of Chemicals (GHS), would incorporate scientific advances regarding persistence and in bioaccumulation, and would have increased scientific rigor (ACC, 2011a).

The concept behind the tool is of a matrix with hazard on the vertical axis and exposure on the horizontal (see Figure 5-5). The hazard level is classified as low, medium, medium high, or high, using the same criteria used by the GHS. The hazard is scored separately for human hazard and environmental hazard, and the final hazard ranking is based on the higher of the two. If existing sources of information on a chemical provide insufficient information to determine the hazard level, the highest hazard score is assigned.

The exposure ranking for a chemical is determined by adding scores from three components: the use pattern, production volume, and persistence and bioaccumulation. The use pattern score is derived from the Chemical Data Reporting Rule, the rule by which, under TSCA, companies report periodically on substances that are currently in commerce. Substances used by consumers get a score of 4, those in commercial use get a 3, those in industrial use get a 2, and those in intermediate use get a 1. Similarly, production volume scores are 4 (more than 100 million pounds), 3 (1 million to 100 million pounds), 2 (25,000 to 1 million pounds), and 1 (less than 25,000 pounds). The persistence and bioaccumulation scores are 5 for substances that are both persistent and bioaccumulative, 3 for those that are one but not the other, and 1 for those that are neither persistent nor bioaccumulative. The three scores are added together to get the total exposure score, which is then used to assign substances to an exposure band or range: low (3 or 4), medium low (5 or 6), medium (7 or 8), medium high (9 or 10), and high (11 or 13).

---

[12] Further information on ACC's approach to prioritization is available at http://www.americanchemistry.com/TSCA (accessed April 2, 2014).

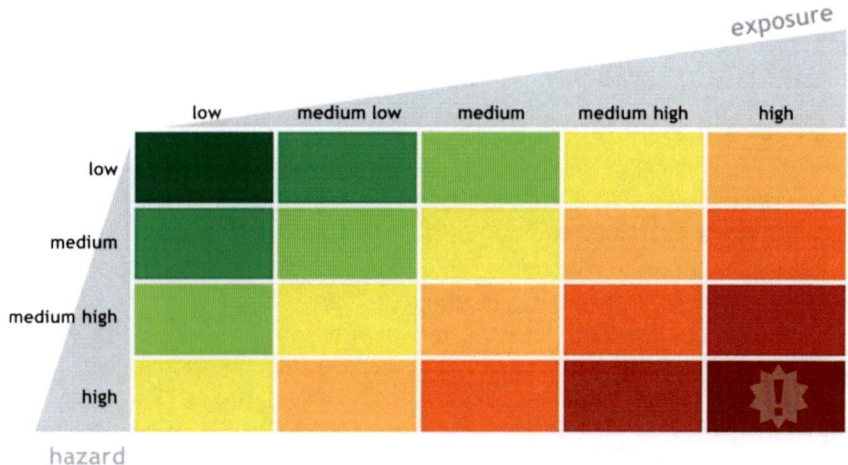

FIGURE 5-5 Concept behind prioritization tool.
SOURCE: ACC, 2011b. Reprinted with permission from the ACC.

The exposure ranking for a chemical is determined by adding scores from three components: the use pattern, production volume, and persistence and bioaccumulation. The use pattern score is derived from the Chemical Data Reporting Rule, the rule by which, under TSCA, companies report periodically on substances that are currently in commerce. Substances used by consumers get a score of 4, those in commercial use get a 3, those in industrial use get a 2, and those in intermediate use get a 1. Similarly, production volume scores are 4 (more than 100 million pounds), 3 (1 million to 100 million pounds), 2 (25,000 to 1 million pounds), and 1 (less than 25,000 pounds). The persistence and bioaccumulation scores are 5 for substances that are both persistent and bioaccumulative, 3 for those that are one but not the other, and 1 for those that are neither persistent nor bioaccumulative. The three scores are added together to get the total exposure score, which is then used to assign substances to an exposure band or range: low (3 or 4), medium low (5 or 6), medium (7 or 8), medium high (9 or 10), and high (11 or 13).

Substances are then placed on the prioritization matrix according to their hazard and exposure scores, with a total prioritization ranking calculated by adding up the hazard ranking (1 through 4) and the exposure ranking (1 through 5). The total prioritization ranking is thus a number between 2 and 9, from lowest hazard/lowest exposure (Priority Group 2) to highest hazard/highest exposure (Priority Group 9). The intention, Franz said, is that an agency should focus on substances in Priority Group 9 first as being the highest priority for further evaluation.

There is also a second tier to the prioritization process that allows for more qualitative scientific judgment, she added. The qualitative factors can be used to move a substance up or down in priority within a given priority group and thus create a rank order within a priority group. The factors considered here are biomonitoring, whether a substance is used in a children's product, emissions information, and whether there are any international risk management actions pending on a particular substance. "If you are in Priority Grouping 9, any one or more of these considerations would, perhaps, move you to number one in Priority Group 9," she said.

In conclusion, Franz offered a number of benefits that using the ACC prioritization tool provides:

- The prioritization tool is based on objective scientific criteria regarding both hazard and exposure.
- It addresses human health and environmental safety.
- It is transparent and offers a helpful matrix visualization.
- The highest priorities are very clear.
- The exposure indicators used are use, volume, and persistence/bioaccumulation. The final indicator incorporates scientific advances regarding persistence and bioaccumulation.
- The tool is flexible and can be updated to accommodate improved scientific information.
- The qualitative factors can be used to influence rank ordering within a priority grouping.

DISCUSSION

Lynn Goldman asked a general question of the panel members whether they heard things from each other that they could apply in the own areas, particularly concerning the issue of prioritization. Richard Judson replied that his group, which consists mainly of chemists and biologists, started from a hazard-centric standpoint, but it has spent a great deal of time over the previous 6 to 8 years talking with the Canadian group. Betty Meek of that group had recommended that instead of initially focusing on hazard, they start on the exposure side. "We felt that our high-throughput hazard approach would let us set priorities first, which we could then refine with lower-throughput exposure estimates," Judson said, "but now we have found ways to do the initial exposure estimates computationally for most chemicals, while we can't currently run the hazard screens for the complete set of tens of thousands of chemicals. There is some low hanging fruit on the exposure side." The lesson he has learned is that it is important to move back and forth between different viewpoints. "You do as much exposure as you can, using inexpensive models. Then do as much high-throughput hazard measurement as possible. You go back and forth, which is a little bit like what the ACC tool is doing. You are almost prioritizing prioritization."

Gina Solomon answered that the different approaches are not mutually exclusive and that having multiple approaches to determining priorities can be beneficial because the priorities being set might depend on context—not just the regulatory, but also whether the prioritization is at a community or statewide or national level and whether one is prioritizing for screening or testing or risk assessment or risk management. "I like multiple approaches to priority setting," she said. "Let's do several different things and hope that we can then elevate the things that should be elevated."

Luz Claudio, from Mount Sinai School of Medicine and a Roundtable member, commented that many in vitro models have been around for more than 20 years and asked Richard Judson, "Are we any closer to a clear guideline of how to use in vitro testing—whether for screening or just to make a dent in the big list of chemicals that have not even been tested in any way?"

Judson began by noting that there are old in vitro tests and new in vitro tests. "Most of what we are using was developed in the last 5 to 10 years, and there are new ones becoming available all the time." These new tests are more focused on molecular targets—molecules that are known, when triggered, to lead to certain types of phenotypic changes. "Then there are intermediate phenotypes," he continued. "Chemicals cause whole-cell changes. Many of the new systems have multiple cell types in the same well, which allows us to see the effects of cell-cell communication. It is getting more and more complex. For instance, some of the new systems have xenobiotic metabolism active within the well." And there is a lot of new technology coming online, he added. "By running lots of assays and combining them computationally, we believe we are getting a better picture of what would go on in a whole animal."

Judson then noted that a major issue in the regulatory area is to validate such tests. One of the reasons that so many old tests are still being used, he said, is that even for simple tests it can take 5–10 years to go through the validation process. "There are other realms of medical testing where that would be crazy, but, somehow, the toxicity testing area has allowed this approach that, oh, a test has to be perfect. . . . That has made it difficult to even imagine how we would move these new tests into the regulatory area." His group has suggested the formation of an international group that could come up with ways to speed up the validation process, he said. Until that happens, there will be great uncertainty concerning just when the new tests will start being used.

Dennis Devlin, from ExxonMobil Corporation and a Roundtable member, asked Judson about the ToxCast project. "I think many of us, if not all of us, were disappointed that ToxCast didn't show more predictive power, at least currently with the information that we have," Devlin said. Then he asked Judson if he thought that these in vitro assays could be used credibly for questions that need relatively immediate answers. Devlin had heard, for example, that the ToxCast assays had been used to screen the dispersants that had been used in the Deepwater Horizon spill, and he wondered how that could have been done credibly. Devlin had also heard it suggested that ToxCast assays could be used to screen the hydraulic fracturing fluid used in wells. Again, he wondered if that could be done credibly, given that there are thousands of sites and every well will have a different mixture. In such cases, Devlin noted, it is assumed that there will have been some exposure, so he asked Judson, "Do you envision a time where the assays will be able to do that for just a hazard assessment?"

Judson noted that there have been two or three talks and papers that have said ToxCast failed. "We are a big science program. We are a target for people." Still, he said, it is important for him and his colleagues to understand the criticism and to take it seriously. There are three ways in which the program might fail, Judson said: The data could be wrong, there aren't enough assays or the right assays, or the models are not sufficient.

The data are certainly not perfect, Judson acknowledged. "Most of the critiques are saying, you have failed because you can't exactly replicate in vivo rodent animal studies," he said. One of the things that has not been included in the work is pharmacokinetics data. "We are just now getting enough of that data," he said. "If you don't have pharmacokinetics, that is a huge driver of a chemical's potential to be toxic. You really have to get the dosing right. We have started to show that if you put the toxicokinetics there, you improve the estimates. And finally, we know that we don't have all the right pathways assayed."

The more important issue, Judson said, is that the biology is really complicated. Effects are not at all linear. "By including more biology, we will get there," he said. "That is not today or tomorrow. I think we will get there just because I believe that biology is not magic."

As for the specific issue of the use of ToxCast for screening, Judson explained that his group had been asked a very specific question concerning the Deepwater Horizon spill. "There were seven or eight dispersants," he said. The exact ingredients in them were confidential business information, but it was believed that some of them contained nonylphenol ethoxylates, which will degrade to nonylphenol, which is an estrogenic. The worry was that if a large amount of estrogenic material was deposited into the shrimp breeding grounds along the Gulf Coast, there would be a population crash.

"The manufacturers wouldn't publicly say what was in there," Judson said. "The idea was, we will just take some of those dispersants and test whether they are estrogenic or not. We did that. It turned out there were two [substances] in which we got a small signal. The one that was approved to be used didn't [have a signal]." Later, he said, through sleuthing through the Internet his group found out that the analysis did get the right answer. Thus, ToxCast was successfully used to answer a specific question.

As for fracking, Judson said, his group has not been asked to do such an analysis yet, but they could. "A similar experiment we have been asked to do is take river water samples downstream of an effluent, which are complicated mixtures, and ask, is there anything that looks estrogenic or androgenic or that is hitting the aryl hydrocarbon receptor?" With their high-throughput screens they have been able to successfully do the same sorts of analyses that have traditionally been done by more complicated and more expensive in vitro screening. "We can answer that sort of question," he said.

REFERENCES

ACC (American Chemistry Council). 2011a. *ACC prioritization screening approach.* Available at http://www.americanchemistry.com/Prioritization-Document (accessed April 2, 2014).

ACC. 2011b. *PowerPoint presentation on ACC's prioritization system.* Available at http://www.americanchemistry.com/Policy/Chemical-Safety/ TSCA/PowerPoint-Presentation-on-ACCs-Prioritization-System.pdf (accessed January 28, 2014).

California EPA and OEHHA (California Environmental Protection Agency and Office of Environmental Health Hazard Assessment). 2013. *California communities environmental health screening tool, version 1.1 (Cal-EnviroScreen 1.1): Guidance and screening tool.* Available at http://oehha. ca.gov/ej/pdf/CalEnviroscreenVer11report.pdf (accessed April 2, 2014).

California OEHHA. 2014. *CalEnviroScreen 1.1 results: Highest scoring ZIP codes with CalEnvironScreen 1.1 scores.* Available at http://oehha.maps.arcgis. com/apps/OnePane/basicviewer/index.html?appid=5e1542837d4246b282dd baa92b0e790f (accessed April 4, 2014).

IUPAC (International Union of Pure and Applied Chemistry). 2007. Glossary of terms used in *Toxicology, 2nd edition. Pure and Applied Chemistry* 79(7):1153–1344. Available at http://sis.nlm.nih.gov/enviro/iupacglossary/ glossarym.html (accessed April 2, 2014).

Knudsen, T. B., and N. C. Kleinstreuer. 2011. Disruption of embryonic vascular development in predictive toxicology. *Birth Defects Research, Part C: Embryo Today* 93(4):312–323.

Patterson. H. 2013. *Health Canada's experience with existing substances under the Canadian Environmental Protection Act.* Presentation at the Institute of Medicine Workshop on the Identifying and Reducing Environmental Health Risks of Chemicals in Our Society, Washington, DC.

Zang, Q., D. M. Rotroff, and R. S. Judson. 2013. Binary classification of a large collection of environmental chemicals from estrogen receptor assays by quantitative structure-activity relationship and machine learning methods. *Journal of Chemical Information and Modeling* 52(12):3244–3261.

6

Current Efforts to Reduce the Risk of Chemicals in Our Society

The workshop's fifth session was devoted to a variety of approaches that institutions have taken to reduce chemical risks. As session chair Al McGartland, Director of the National Center for Environmental Economics at the U.S. Environmental Protection Agency (EPA), noted in his opening remarks, the session's seven speakers have two things in common. First, they are all leaders in green chemistry, sustainability, or related fields. And, second, they all have broad familiarity with both the scientific aspects and the institutional aspects of finding and implementing solutions to problems related to chemical risks. It seems likely, McGartland commented, that success requires both scientific and institutional competence.

CASE STUDY: SUSTAINABILITY AND GREEN PROGRAMS AT THE NATIONAL INSTITUTE OF ENVIRONMENTAL HEALTH SCIENCES

In the session's first presentation, Trisha Castranio, Sustainability Analyst at the National Institute of Environmental Health Sciences (NIEHS), described the sustainability, eco-friendly, and green business practices at NIEHS. Castranio, who develops sustainability policies and environmental management goals for NIEHS and is responsible for evaluating the effectiveness of its stewardship initiatives, offered a detailed accounting of exactly how the institute works to reduce the amount of harmful chemicals that it uses and disposes of.

She began with a brief description of NIEHS. It is a biomedical research facility. Its intramural program has more than 100 groups working onsite, while its extramural program operates 17 different programs and centers doing work in disease research and exposure research. Both the intramural and extramural programs are very productive and regularly contribute to the peer-reviewed literature.

The institute places a great deal of emphasis on sustainability issues and green business practices, and it has had success in those areas, Castranio said. "We have been awarded Green Championship awards from the Department of Health and Human Services for 3 of the last 4 years," she reported. The institute also received one award for sustainability reporting and another for environmental stewardship for its composting program.

Much of what the NIEHS has done over the past 5 years to move in the direction of sustainability and green practices has been harvesting "low-hanging fruit," Castranio said. The initiatives include such things as installing the more energy-efficient light-emitting diode (LED) lighting, composting food wastes from the cafeteria, and encouraging more recycling. One initiative, whose goal was to reduce energy use and thus the production of greenhouse gases, replaced old, less-efficient ultra-low-temperature freezers with newer, much more efficient freezers. Each of the older freezers used as much energy as an 1,800-square-foot home, Castranio said. "This project was a great way to not only reduce the carbon footprint, but it also got people to go through their freezers," she said. Because laboratories had to get rid of two old freezers to get one new freezer, the researchers had an incentive to get rid of items that had been in their freezers but were no longer needed.

NIEHS has reduced water usage by 35 percent over the past few years, Castranio said, and it has reduced energy use in various ways, from overnight shutdowns for the information technology groups to changing temperature settings in the buildings to require less heating or cooling.

One of the most challenging issues has been dealing with the many different chemicals that are used in the NIEHS laboratories. The researchers there work with a wide variety of chemicals and reagents. Their jobs demand it, and they are generally very well organized and careful in their handling of the chemicals. But they tend to have a lot of chemicals—often far more than they need. A group of researchers may have been working in the same laboratories for 30 years, with postdoctoral students coming in for a few years and leaving again, and it is natural to want to save the different materials that have been used in various experiments over the years, Castranio said. "People think, I am saving this for this person, and somebody might need that and the next person might work on that. And then these tend to build up. That belongs to somebody else. They are not going to get rid of it. Nobody wants to touch it. That kind of thing."

Part of her job is to reduce the amount of chemicals in the laboratories. The first step is to get researchers to take a careful look at what they have and create an inventory. That allows Castranio to keep track of usage, which is the first step in developing plans to reduce the amount of chemicals the laboratory is using and to attempt to move to chemicals that are greener. Of course, the first requirement is that the researchers must be able to use the materials that they need in order to do their work effectively, but there will be some materials that researchers can use less of or can replace with something greener. "They are going to have some things that affect them directly and other things that do not," she said. "There will be some give and take. It is going to be on an individual basis."

In looking to reduce the laboratory's use of toxic and harmful chemicals, it was important to not overlook common spaces, Castranio. Equipment rooms and storage rooms tend to become catchall spaces. Researchers will buy large quantities of various materials and place them there. The cold rooms also end up being catchall spaces whether they require cold storage or not. All of these spaces require a walkthrough to see what is there and what can be disposed of.

One key to using fewer and greener chemicals is simply to get the researchers thinking about the chemicals from a life-cycle perspective. "Once we can get people to think about where their chemicals are going, where they come from, what is going to happen to them after that, then we can get them to reduce and possibly find a new way to do that type of research," she said.

> "The most important thing for me is to reveal the face behind the bottle. That means . . . how to handle this chemical or how to handle that potential waste. Before you, somebody harvested it or somebody synthesized it. Somebody packaged it. Somebody shipped it. Somebody brought it to you. Somebody put it on your desk. Now it is on your desk. Then where does it go? Who is going to handle it? Who picks it up? Who takes it there? Where does it go after that? These are the things that I think will help people think twice about how much they use."

On the operational side, Castranio said, one of the most important things is making sure that environmental considerations are part of the planning from the inception of a project and do not come in as an afterthought. Waiting until the project is 90 percent complete never works because "there is no extra money, and green almost always means more money," she said. "It has to get in at the beginning, and then you can have the trade-offs."

Such thinking is more difficult with research projects, she noted. "I had some people come to me and want to talk about 'greening the grants.' Grants are merit-based scientific funding. We cannot really choose funding based on how green their process is." However, she added, it is possible to talk to the scientists and have them make efforts to keep their projects as green as possible.

From her experience at NIEHS, she offered advice for anyone who wishes to institute programs to reduce the amount of chemicals used by society. It starts with measuring and reporting, for it is vital to know what one is dealing with. Once you have a clear baseline, you begin to set goals, implement programs to reduce chemicals, and make sure that best practices are shared widely.

The scientists at NIEHS have a dual purpose, she said. They do scientific experiments that relate to toxicology, but they also should be doing them in the least toxic way. "We are an environmental institute. We should be doing this."

CASE STUDY: JOHNSON & JOHNSON

The session's second speaker was Zephanie Jordan, Vice President of Global Regulatory Affairs and Product Stewardship at Johnson & Johnson. To put her talk in context, Jordan explained that Johnson & Johnson has three major divisions—the pharmaceutical products division, the medical devices division, and a consumer products division—and that she works in the consumer products division.

The goal of Johnson & Johnson's sustainability initiative, which was launched in 2011, is summed up by the slogan, "Caring for a healthy future." In particular, the company's sustainability initiative has three main aims: to promote healthy people and communities; to promote a healthy planet, minimizing waste and conserving finite resources; and to promote healthy business, which the company believes will follow naturally from focusing on healthy people and a healthy planet. "What we mean by healthy business," Jordan explained, "is that the most trusted brands will thrive and endure."

A key to building this sort of consumer trust, she said, is ensuring transparency "into what we do and how we do it." In August 2012 the company launched its Safety and Care Commitment, which Jordan characterized as an effort to provide better transparency into the safety assurance processes that Johnson & Johnson uses in the development of its products and also into the policies governing ingredients in various products, such as beauty care products and baby care products, where concerns about safety are particularly acute.

The company has a five-level safety assurance process, Jordan explained. The five levels are sourcing raw materials, toxicology assessment, clinical evaluation, in-use testing, and continuing evaluation.

Johnson & Johnson's background in pharmaceuticals shapes its approach to sourcing materials, Jordan said. "That background in health care causes us to look first through a safety lens," she said. "We have very high standards and sustainability principles for our ingredients. We only partner with suppliers that can meet those standards." To illustrate, she noted that the company requires each of its suppliers to fill out a 12-category questionnaire covering a wide range of topics. "We were one of the first companies to move to next-generation ingredient reviews where we require all of our suppliers to disclose compositional information down to one part per million," she said. Because the general standard is in the range of 100 to 1,000 parts per million, the company believes its approach to be at least 100 times more sensitive. For each of its suppliers Johnson & Johnson requires independent certification of various aspects of the firm's operations, from conditions on its production floor to its business practices before it will partner with that firm.

Second, each ingredient that Johnson & Johnson uses must pass a toxicology assessment. The first step in that assessment is for the company's global team of toxicologists to assess the data that are available. They look at both the hazard and the risk that an ingredient might pose. "We meet or exceed regulatory standards for all of our materials," Jordan said. The company also examines how an ingredient is going to be used. "Is it likely to be used in a shampoo that gets washed off? Is it likely to be used in a lotion that gets left on the skin?" The company uses that information to put the safety assessment in context.

Next the company carries out a clinical evaluation for each of the formulations that it develops, looking at both efficacy and safety. In particular, the evaluations assess the safety of specific concentrations and ingredient formulations. The products are tested for such things as irritation, sensitivity, and response to sunlight.

Because it is not enough to look at the products in a controlled clinical setting, before a product goes to market the company will also put it into user testing. "We have a bank of volunteers around the world that take the products into their homes and use them," Jordan said. "We are looking for unanticipated ways that consumers or people may use our products so that we can adjust the formulation or the label information to ensure that they are going to be safely and effectively used in the home."

The last level is the continuing evaluation. "We talk about this as step five," she said, "but it underpins the whole process. It never stops. We are always evaluating our products." The company has an ingredients working group made up of scientists and medical professionals from across the globe that is constantly reviewing new and emerging data. They are evaluating three things, Jordan said. First is the science and what is emerging in terms of the research concerning the ingredients the company uses. Second is regulatory trends, and third is consumer attitudes. "They are looking at what consumer sentiment is telling us about what we ought to do with our ingredients," she said. From that the working group makes recommendations on product formulations, labeling, packaging, and instructions for use and these are taken up in our internal policies.

The working group has a good track record, Jordan said. "Typically, we find that this group has made recommendations and we have implemented those into our ingredient usage policies years before there have been regulated controls on ingredients."

To improve the sustainability of its products, Johnson & Johnson instituted its Earthwards program, which is an internal certification program for its products. As company researchers are developing a new product, they must consider seven areas in terms of sustainability: materials, packaging, energy, waste, water, social, and innovation. "Each product is scored against these," Jordan said, and a product is awarded Earthwards certification if it shows significant improvement over existing products in at least three of the seven areas. "We do not have to achieve certification for every single product," she said, "but we have to score against these. It is intended to ensure our scientists look for improvements in every one of these areas."

A second voluntary program that the company instituted is called Global Aquatic Ingredient Assessment, or GAIA. It is a tool that the company uses internally to assess the impact that an ingredient or formulation might have on the environment. In particular, the company scores a product on three measures: persistence, bioaccumulation, and toxicity.

As part of its Safety and Care Commitment, the company is making public some of its internal ingredient standards in an effort to put itself in a leadership position on certain products of concern. "Essentially, we made some commitments to remove certain ingredients or trace materials or restrict their use in the categories of products," Jordan said. "We did not necessarily do this for safety reasons. We did it because consumer perception and sentiment was such that even if we continued to use some of these ingredients that are safe, they were not acceptable from a consumer perception perspective."

As an example, she described the company's position on formaldehyde-releasing preservatives. These chemicals are generally very effective and very safe, Jordan said. There is, for instance, 14 times more formaldehyde in an apple than in a bottle of Johnson & Johnson's baby shampoo. "You would need to bathe a baby 40 million times in 1 day to achieve the California Proposition 65 level for labeling." Nonetheless, there is enormous public pressure to move away from these preservatives, and so Johnson & Johnson set a goal of removing formaldehyde-releasing preservatives from all of its baby products by December 2013 and decided not to use them in any new adult products unless a special exception is granted.

The company is walking a fine line here, Jordan said. "We have to be very careful about unintended consequences when we do things like this. . . . When we are taking something out of our products, we have to be very sure that what we are replacing it with is going to be suitable and appropriate and that we do not see unintended consequences in other ways." For example, because of public pressure, the industry is moving away from some preservatives to others, and increased exposure to these preservatives has led to an increase in sensitization rates in one instance.

The company's reformulation work is a "mammoth task," she said. "We are reformulating around 200 products over the course of a couple of years. It takes us 18 months to 2-and-a-half years to reformulate a single product because we go through that five-level safety assurance process for every product that we formulate. We are doing this on a global scale. We have diverted resources to this effort because this is considered to be a high priority for the organization."

In closing, Jordan offered two parting thoughts. First, she said, it is vitally important that decisions about chemicals and ingredients be based on data and that the decisions be put into context with a risk-based approach. Second, it is crucial how information about these products is transmitted to the public. "We know that people that use our products want helpful information and they want assurance of safety," she said. "There is mass confusion in the society about these ingredients. We have been trying to work out how to solve that part of the problem. Our thinking is moving toward less about providing details on ingredients that can be misinterpreted and more in terms of driving behavior change and helping consumers ask the right questions as they choose products for themselves and their families."

CASE STUDY: DOW CHEMICAL COMPANY

The next speaker was Connie Deford, Director of Global Products, Sustainability, and Compliance at Dow Chemical Company. She is responsible for leading Dow's global product sustainability program, and she described those sustainability efforts to the workshop audience.

Dow was founded in 1897 by Herbert Henry Dow in Midland, Michigan. Dow chose that particular location, Deford explained, because of the presence of brine wells, which served as a source of chlorine and caustic soda, which were important raw materials for the new chemical company. Today the company has annual sales of $60 billion and has 188 manufacturing sites in 36 countries. It supplies plastics and chemical products to customers in 160 countries.

Deford began her discussion of Dow's sustainability program by talking about motivations. "Environmental health and safety and sustainability are at the core of what our company is all about," she said. There are a number of specific motivations, including the local protection of human health and the environment, addressing climate change, encouraging energy efficiency and conservation, product safety leadership, and contributing to community success.

"Certainly, our customers and our customers' customers are key drivers for many of our programs and activities in the sustainability space," she added. "Most of you are probably very familiar with Walmart's support of The Sustainability Consortia. Most recently, Target has introduced their sustainable product standard. Clearly, we listen to those retailer activities. These retailers' actions are motivators for the kind of work that we are doing." The company also pays attention to regulations and green certification programs "to signal where we should spend our time and energy and focus our innovation efforts."

To identify and address sustainability issues in its existing products, Dow takes a multipronged approach, Deford said. One key tool is the use of life-cycle assessments. These assessments go beyond simply looking for potential hazards. "We look at what type of waste might be generated and emission from our manufacturing processes," she said. "We look at what happens at the end of the life cycle. Are there things that we can do working with our customers to improve their utilization of our products? These are key ways in which we identify opportunities to look at addressing gaps in the sustainability space for our existing product portfolio."

The company also assesses products against its 2015 sustainable chemistry goal criteria.[1] In its efforts to meet these goals the company looks at a variety of criteria for its products. "We start at the very beginning—the raw material extraction point. What can we do? Is there more that we can do relative to sourcing of raw materials? Can we do more recycling within our operations?" In addition to looking for ways to use more renewable and recyclable materials, the company also looks for ways to make its manufacturing process more efficient and ways to use less energy in transporting the products. "We partner closely with our supply chain organization, looking at opportunities to relocate facilities nearer our facilities." They also examine the various ways their products are used, looking for opportunities to move in the direction of greater sustainability.

To strengthen its product safety program—which is another prong of the sustainability effort—the company uses the prioritization tool developed by the American Chemistry Council (and described in the presentation by Christina Franz in Chapter 4). "We look critically at products that are going into consumer applications, as well as products that might have a higher degree of hazard," she said. "We do not look at every chemical and application the same."

Another aspect of strengthening the safety program is a critical look at the company's ingredient disclosure practices. "We are listening to our customers and customers' customers asking and wanting to know more about those materials that are in the products that we supply to them," she said. "We are challenging our businesses to look critically at how important is it to maintain that confidentiality." It is important to find the

---

[1] Further information on Dow's Sustainability Goals is available at http://www.dow.com/sustainability/goals (accessed April 2, 2014).

right balance between providing the product information that customers want and need to know versus not providing too much product information to competitors and thus losing competitive advantage. "We are continually challenging our businesses about how to do a better job in that regard."

To illustrate Dow's approach to sustainability and some of the challenges in such an endeavor, Deford offered two case studies. The first concerned the search for alternatives to nonylphenol ethoxylates (NPEs). These chemicals have historically been used as surfactants. They continue to be widely used in applications where their exposure to the environment can be minimized, but the company was looking for viable alternatives for uses in which the chemical may be released into the environment.

"There are lots of surfactants out there," she said. "It was not an issue to find a surfactant to replace NPEs. The challenge has always been finding a surfactant that is cost effective and stable in the environment that it needs to be used with a price that is similar to NPEs. NPEs are an very cost-effective surfactants."

The company developed a new line of surfactants, the ECOSURF EH Surfactants. Readily biodegradable and with low aquatic toxicity, they were designed to help formulators meet rising expectations for performance and convenience, Deford said. In particular, they can help formulators comply with regulations and more stringent health and environmental certification programs.

It took significant investment to develop those new surfactants, she said. The company not only had to design new chemistry, but it had to gather data to help customers understand how it would perform in different applications as well as data regarding the new chemicals' health and environmental implications. Dow also had to make the capital investment to reconfigure a manufacturing facility to make the new surfactants.

Unfortunately, the cost of the new chemicals has resulted in limited success even though tests showed them to be particularly effective in certain agricultural formations. "It demonstrates that although it is a very effective technology, it also has to be cost effective."

The second case study Deford offered concerned the development of a polymeric flame retardant to replace hexabromocyclododecane. HBCD, as it is known, has been used as a flame retardant for a long time in a variety of different applications. One of those applications is use in extruded polystyrene (XPS) foam, such as the STYROFOAM® brand rigid insulation that Dow manufactures. Dow researchers played a key role, Deford said, in the development of polymeric flame retardant, or Poly FR, as an alternative to HBCD as a flame retardant in XPS foam. Poly FR is a large-molecular-weight material that does not have the potential to bioaccumulate, and it does not have the same toxicity concerns as are associated with HBCD, Deford said. Recently the Design for the Environment program at the EPA released a draft report that found Poly FR to be a viable alternative to HBCD and that predicted that Poly FR would be safer than HBCD.

The development of Poly FR required an even greater investment than the development of the ECOSURF EH surfactants, Deford said. "There were many years of effort expended in looking at an alternative to HBCD," she said. "We screened commercially available products, but in the end, this was a unique chemistry that was identified to replace the material." Significant time, effort, and expense went into development and laboratory testing as well as conducting product certification testing to confirm performance. Adjustments also were required in formulations. Poly FR is clearly a success story, she said, but it is a success story that illustrates just how much time and energy and investment it can take to replace a successful product.

For that reason it is crucial that Dow be able to learn about the environment, health, and safety potential of new products as quickly as possible. The company is committed to bringing products to market that are safer and more sustainable than the products they have been selling in the past, but they need to be able to predict with some accuracy which potential products are likely to be successful.

Dow is beginning to utilize its predictive toxicology testing to meet its information needs during product development. The company has been working with the EPA, academic researchers, and others to optimize such testing. Dow also uses a life-cycle assessment to determine how sustainable a new product is likely to be. And, Deford said, the company uses "a very rigorous stage gate process so that at various stages we are asking ourselves those right questions about whether or not a material is more sustainable than the material that we are replacing." In particular, Dow judges each new material on six dimensions of sustainability: economic, social, greenhouse gas emissions, water, resource use (energy and raw materials), and the Dow 2015 Goals Composite. If at any stage the new material does not seem to be a significant improvement, the company will not move on to the next stage of development.

CASE STUDY: AMERICAN CHEMICAL SOCIETY GREEN CHEMISTRY INSTITUTE

David Constable, Director of the American Chemical Society's (ACS's) Green Chemistry Institute, spoke next. In his presentation he made the case that innovating toward sustainability in chemistry is a business imperative.

He began by drawing a distinction between end products becoming greener and the chemical building blocks that are used to create those end products becoming greener. Both tasks are important, but the latter may be more difficult.

The challenge, Constable said, is that chemists must work with what is available, "and the basic building blocks and what is at hand for your average chemist is inherently hazardous and toxic. The chemicals and chemistries that chemists use are based in technologies that are 150-plus years old, by and large. They are inherently using toxic materials. Chemists are creatures of habit, and they use the same things over and over again."

But it is not just a matter of habit, he said. The chemical reactions and processes that chemists rely on today have been refined by decades of experimentation and improvement. They are easily obtained at low cost, they react in predictable ways, they have been optimized to produce maximum yields, they take place via thermodynamically and kinetically favorable reactions, and they generally do not require sophisticated reactors or technology in the laboratory. On the negative side of the equation, however, the current chemical building blocks have a variety of sustainability risks. Their feedstocks are sometimes hazardous to human health or the environment, the intermediate materials in the processes can also be hazardous, the chemical processes that are used can be high risk, and there can be inappropriate engineering or process controls. Thus, despite the advantages of today's chemical building blocks, there are a variety of reasons to look for greener alternatives.

However, Constable said, it will not be easy to change behaviors and preferences in the chemists who do this work. "The entire system is likely to have to change."

First, the types of chemicals that chemists work with will need to be changed. "Chemists do what they do because the molecules that they are working with are reactive molecules by definition," he said. In other words, chemists tend to choose chemicals that can be put together in a reaction vessel and naturally react with one another, with no additional work required from the chemist. They generally do not think in terms of reactions that are not quite so natural but that can be produced under the proper conditions. "The point is the entire way in which chemists are educated . . . does not necessarily get them to the place that we need them to be."

Chemists also tend to think in terms of maximizing yield, Constable said, but that is not a good measure for green chemistry, in which other factors are more important. Chemists like to use chemicals that are familiar and easily obtained. "Chemists reach for the same solvents that they have used for years. They use the same framework molecules. They go to the same purveyors of chemicals, and that is all they use." Green chemistry will require that they leave this comfort zone.

Given this inertia, what steps could encourage the sort of innovation that will be necessary to move to green chemistry? Constable first mentioned the regulatory option. "If you regulate something out of existence, people do not have any choice but to find something different," he said. "It is not really the preferred way to do it, but it will get people to think about it."

But what many supporters of green chemistry foresee is that it will eventually be possible to develop green chemistry to the point that it will naturally edge out traditional chemistry. This will require green chemistry to hit what Constable described as the "sweet spot": green chemistry that is environmentally preferred to traditional chemistry, is economically viable, and offers equal or better performance.

There are a variety of reasons for businesses to move to green chemistry, he said. For example, it can cut costs in various ways. "If you are not having to deal with managing toxic chemicals, it is inherently cheaper," he said. The biggest cost of running a laboratory is the ventilation through the hoods, not the chemistry itself, so if it were possible to do chemistry in a way that did not require hoods, that would lead to a substantial decrease in energy costs. There are also a number of intangibles related to how customers view green products versus traditional products. For example, he said, people are very concerned about the toxic chemicals in the supply chain, even if the finished products are not hazardous. Green chemistry could remove that problem.

One of the ways the ACS Green Chemistry Institute has been trying to encourage the development and use of green chemistry is through the creation of industry roundtables. "We have asked companies to get together and to talk about issues in a collaborative and noncompetitive way." The purpose of the roundtables is to address technical challenges, develop decision-making tools, inform the research agenda, drive the use of good science in setting policy, and influence the adoption of green chemistry throughout the supply chain.

One of the three roundtables that the ACS Green Chemistry Institute has established is a pharmaceutical industry roundtable, which started up in 2005. This sort of cooperation among pharmaceutical companies makes sense, Constable said, because the companies compete on the basis of active pharmaceutical ingredients, not the chemicals and chemistries by which the products are made. "There is an awful lot of room to talk about how to get toxics out of products and processes without affecting what they actually compete on," he said.

Two other roundtables have been established within the past 5 years, a chemical manufacturer's roundtable and a formulators' roundtable. It has been difficult to get the major chemical manufacturers to the table, Constable said. "The reason for that is largely because they compete at lower margins, and they compete in some of the same product areas. Also they are more diversified." Still, he said, there are areas in which chemical manufacturers may be able to cooperate, such as alternatives to distillation. In the manufacturing of chemical products, distillation processes account for more than 25 percent of the energy use. "If we can come up with alternatives to distillation, it will drive a lot of very good behaviors and outcomes from a green chemistry standpoint."

Another way to improve sustainability would be to develop new catalysts. Catalysis is used somewhere in the supply chain for probably 40 to 50 percent of the chemicals on the market, Constable said. "It is used everywhere, and everything that you use on a daily basis can usually be traced back to some catalytic process." The major metals of concern that are used in catalysis are platinum-group metals, which include metals like platinum, palladium, rhodium, ruthenium, osmium, and iridium. What many people do not realize is that mining for these metals results in the release of a large number of toxic materials into the environment. "That is something that is not really perceived at the bench level of a chemist who is choosing a catalyst," Constable said. But some of the more abundant metals whose mining is not so damaging to the environment can also serve as effective catalysts, or in many cases organic enzymes can serve as catalysts. So there are ways of making catalysis greener, but they require looking beyond the immediate chemical processes to think about the entire process, including the costs to the environment of mining the metals used as catalysts.

There are also a large number of ways to move toward sustainability by engineering new chemical processes for use in manufacturing chemical products. For example, Constable said, the pharmaceutical industry today uses mostly batch chemical processing. Moving to bioprocessing and continuous processing would be a way to increase energy efficiency and to reduce the production of toxics and unwanted byproducts. Another area is separation and reaction technologies. These account for an enormous amount of energy use and waste, he said, and simply looking for better solvents hold tremendous potential for reducing the amount of toxics and wastes produced.

Finally, Constable spoke about some of the challenges facing those who are trying to move the industry to greater sustainability. Institutional inertia is a huge problem, he said. Chemical companies are used to making chemicals in a certain way; it is what they know, and it is not easy to get them to take an entirely new approach. A related hurdle is the capital that is invested in doing things the traditional way. Furthermore, when financial analysts examine the costs of doing things the traditional way versus using green chemistry, the traditional measures of analysis they use generally point to the traditional ways of doing chemistry. Many of the ways green chemistry offers a financial advantage, such as in the savings related to sustainability and life-cycle considerations, are generally not considered in traditional measures of profitability.

In addition to such institutional issues there are human behavioral factors that come into play as well. "I would say that a lot of what we talk about boils down to behavioral changes in people at the bench level," Constable said. Furthermore, when senior management is planning for the future, other issues than green chemistry tend to dominate the discussions about sustainable development and corporate social responsibility. Another issue involves the educational system: As long as chemists are taught only the traditional chemistry in school, it will remain difficult to convince them to transition to green chemistry. More generally, people naturally tend to be risk averse and resistant to change. Developing a new system of green chemistry that can replace the existing chemistry will require finding ways to overcome all of these challenges.

CASE STUDY: SUBSTITUTE IT NOW AND GREENSCREEN

The next speaker was Beverley Thorpe, Consulting Co-Director of Communications and Advocacy for Clean Production Action, who spoke about two initiatives to advance safer chemicals in the marketplace. The first was the Substitute It Now, or SIN, List, which was created by the International Chemical Secretariat, based in Sweden. The second was GreenScreen, a comparative chemical hazard assessment method for identifying safer chemical substitutions for chemicals of concern.

Clean Production Action is a small nongovernmental organization (NGO) based in Somerville, Massachusetts. It designs and develops strategic solutions to promote green chemistry products and sustainable materials. It networks with many governments, industry leaders, and other NGOs around the world, Thorpe said. Through that networking it comes into contact with a variety of ideas and methods for moving the economy forward with safer materials.

One such method is the SIN List, which was developed by ChemSec, a nonprofit based in Gothenburg, Sweden, in cooperation with about nine other NGOs around Europe. In following the negotiations that eventually led to the Registration, Evaluation, Authorisation, and Restriction of Chemicals (REACH) legislation, which now governs much of the chemical use in Europe (see Chapter 2), the NGOs came up with the idea of creating a list of the chemicals for which it was most urgent to develop safer substitutes, she explained. That idea became the SIN List.

The SIN List was intended, Thorpe said, to help businesses anticipate which chemicals are likely to be listed on REACH's restricted list as well as to clarify for businesses the criteria that will be used to determine which substances would most likely make the candidate list of chemicals that would then need to seek authorization for continued use in the European market. By making it easier for business to understand how various chemicals would meet the REACH classification for chemicals of concern, Thorpe said, the SIN List could help companies fast-track these for substitution.

REACH came into effect in June 2007, and the first SIN List was created in 2008. Version 2.1 of the SIN List was released in 2013. The development of the REACH restricted list has been a slow ongoing process, and the idea behind the SIN List is to create this list of chemicals of concern quicker. There is no regulatory power behind the SIN List, Thorpe noted, "but it is based on peer-reviewed publicly available data."

Currently, there are about 626 chemicals listed on the SIN List, she said, and they all meet the criteria specified by REACH of being CMRs (carcinogenic, mutagenic, or toxic to reproduction), PBTs (persistent, bioaccumulative, and toxic), vPvB (very persistent and very bioaccumulative), or substances of equivalent concern.

Thorpe suggested that audience members should check out the SIN List. "This is a very interesting database," she said. "It is clear, it is easy to negotiate, and it gives you tons of information." It is possible to search the database using various criteria—such as health impacts, the sector in which the chemical appears, production volume, functional use of the chemical, or registration information such as whether the chemical has been put on the candidate list for authorization. The database also provides the names of producers and the locations of where the chemical is produced.

The SIN List has already affected the chemical use choices of various businesses, Thorpe said. For example, Carrefour, which is Europe's largest retailer, has added SIN to its own list of 600 substances of very high concern, which it sends to its suppliers so that they can begin work on substitution. Sara Lee's Critical Ingredients Program integrates the SIN List, and Skanska, one of the largest construction firms in the world, has integrated SIN into its voluntary restricted substances list.

Furthermore, Thorpe said, investors are also using the SIN List. One research company, MSCI, is applying the list to assess the business risk that companies might be facing in the future as REACH moves forward and chemicals for authorization are listed. In particular, MSCI is using the SIN List to identify the most at-risk product categories and therefore identify the most at-risk companies based on the number of SIN List chemicals they use in their products.

The U.S. Department of Defense is also using the SIN List, Thorpe said. It wants to identify and proactively manage emerging contaminants that can adversely impact human health and the environment, and to better understand the effects that REACH might have on military readiness. Finally, it is using the SIN List as a leading indicator of potential substitutions in commercial off-the-shelf products.

Thorpe next spoke about GreenScreen, a method for advancing informed substitution. "This is a method for comparative chemical hazard assessment developed in house by Dr. Lauren Heine and Dr. Mark Rossi," she said. It builds on the Design for Environment approach of the EPA. It is freely and publicly accessible and can be downloaded from the Clean Production Action website (http://www.greenscreenchemicals.org).

A GreenScreen assessment is done in three steps: (1) assess and clarify hazards, (2) apply the benchmarks, and (3) make informed decisions. For the first step GreenScreen covers 18 hazard end points groups into four categories: Human Health Group I (carcinogenicity, reproductive toxicity, etc.), Human Health Group II (acute toxicity, neurotoxicity, skin sensitization, eye irritation, etc.), Environmental Toxicity and Fate (acute aquatic toxicity, bioaccumulation, etc.), and Physical Hazards (reactivity, flammability).

Once the particular hazards have been identified, the more time-consuming step is applying the benchmarks to the hazard classifications in order to determine how much of a concern a particular chemical is. It is a complex process, but the outcome can be presented as a simple numerical score from 1 to 4, with 1 being "avoid/phase out" and 4 being "inherently low hazard." "It is this benchmarking use of the GreenScreen that companies find useful because you are categorizing chemicals into this kind of continuous improvement," Thorpe said.

As an example of how companies are using GreenScreen, Thorpe described the experience of Hewlett-Packard (HP). HP is interested in GreenScreen, she said, because replacing materials is very expensive. "You do not want to invest in a multi-million-dollar new material and find 2 years later it is going to be restricted within regulation. It makes good economic sense to reduce your business risk and understand what is inherently safer. Plus they want to avoid unintended consequences and identify preferable materials."

HP uses GreenScreen to choose alternatives to substances of concern that must meet GreenScreen benchmark 2 or higher. The company also has found that by articulating material goals to its suppliers, it can really spur innovation, Thorpe said. "Using the GreenScreen benchmarks allows a company to not only tell suppliers what they do not want (e.g., their restricted substances list) but it allows a company to clearly identify the criteria of what they do want."

For example, one of HP's goals is to phase out all halogenated flame retardants and polyvinyl chloride (PVC) polymers. It has integrated GreenScreen into its procurement and specifications to its supply chain. In the specific case of power cords, HP's suppliers must meet GreenScreen benchmark 2 or higher to get on the approved materials list.

The company has now screened well over 30 materials, and several have been approved. The screening is mandatory, Thorpe said, and is in addition to all the usual standard and regulatory requirements. HP requires full disclosure under confidentiality agreements.

"What they find is that the formulators are very engaged," Thorpe said. "They are actively performing GreenScreens. They like it because you are giving them very clear criteria. You are seeing this innovation within the supply chain, which we find really interesting." Thorpe provided a direct quote from one of HP's suppliers, Jonathan Plisco of PolyOne, explaining the advantages of GreenScreen: "The more you know about what you are putting into your products, the more likely you are to make choices in product development."

In the future, Thorpe said, HP will expand the use of GreenScreen to other materials it procures from suppliers, and the company is also helping integrate GreenScreen into the electronic sector generally. Other information technology companies are now using GreenScreen as well, she said, and HP is helping to introduce GreenScreen methodology into ecolabels.

Finally, Thorpe described a set of principles for assessing alternative chemicals that were formulated in 2013 by a group of environmental health scientists, advocates, policy makers, and academics. The Commons Principles for Alternatives Assessment[2] are based on earlier work done by the Lowell Center for Sustainable Production, the Toxics Use Reduction Institute, the Environmental Defense Fund, and the BizNGO Working Group. The aim is to phase out hazardous materials, phase in safer substitutions, and eliminate hazardous chemicals wherever possible. The group settled on six fundamental principles: (1) reduce hazard, (2) minimize exposure, (3) use best available information, (4) require disclosure and transparency, (5) resolve trade-offs, and (6) take action. Or, as Thorpe summarized it, "Our whole modus operandi is to move off of inherently hazardous materials to safer alternatives through informed substitution and more information."

---

[2] The Common Principles for Alternatives Assessment are a set of common definitions and principles for chemicals alternative assessment to be shared and used in framing discussions about alternatives assessment and to guide decision making about safer chemical use. Further information is available at http://www.bizngo.org/alternatives-assessment/commons-principles-alt-assessment (accessed March 31, 2014).

CASE STUDY: MASSACHUSETTS TOXICS USE REDUCTION INSTITUTE

Liz Harriman, Deputy Director of the Toxics Use Reduction Institute (TURI) at the University of Massachusetts, Lowell, spoke next. She discussed the toxics use reduction program and then offered two case studies in finding safer alternatives to toxic chemicals.

The Toxics Use Reduction Act, which established TURI, is intended to sustain and promote the competitive position of Massachusetts industry while promoting a reduction in their use of toxic chemicals. Harriman explained that the law requires businesses to analyze their use of toxic chemicals every other year, to look for opportunities to reduce their toxics use and waste, and to publicly report their toxic chemical use, but it does not require the businesses to implement anything.

TURI has a number of roles that were set out by the Toxics Use Reduction Act. It provides information on toxic chemicals and safer alternatives as well as education, training, and tools for those working on toxics use reduction. It carries out research on and demonstrations of green chemistry and innovative technologies. It provides grants and supports academic research to help connect the needs of businesses with those who have the relevant knowledge about and capacity for reducing the use of toxic materials.

"The decision makers that we are trying to reach are manufacturers, small businesses, community groups," Harriman said. "We need to tailor the information we provide to that audience." On the other hand, she added, "when we go to make decisions on what chemicals should be on the list of the program or should be prioritized, we make a much deeper dive. Our science advisory board requires extensive information to make those kinds of decisions."

TURI is also the science and policy arm of the program. As such it has released a number of reports describing the chemicals used in Massachusetts and various health issues that are known to be related to chemical use. However, she said, the objective of the reports was not to try to establish a direct link between chemical use in the state and cancer rates and other health statistics. "It was really to educate our toxics use reduction planners who were doing those assessments for companies about what the health risks are of the various chemicals they use and to educate the cancer industry and the cancer researchers about what chemicals are being used in industry that might affect cancer rates. It was very much informing each group."

To illustrate the issues that TURI deals with, she offered a case study on perchloroethylene, a chemical widely used in dry cleaning, in industrial vapor degreasing, and in some products for consumers and small businesses, such as brake cleaners. Perc, as it is widely known, is a neurotoxin; is thought to be a human carcinogen; can cause damage to the liver, kidney, and central nervous system; and is toxic to aquatic organisms.

The assessment of alternatives to perc was carried out with a general model developed by TURI in conjunction with the Interstate Chemicals Clearinghouse, IC2 (see Figure 6-1). The seven-step process begins with defining the goal and then identifies the chemicals of high concern. Step 3 is to identify alternatives to those chemicals. These alternatives may be other chemicals or products or perhaps a process that can satisfy the same need as the chemical in question. Step 4 is to prioritize and pre-screen the alternatives, and Step 5 is the alternatives assessment. This involves a technical and performance assessment as well as an environmental, health, and safety assessment, which can be done in various ways—for example, by the GreenScreen tool for a chemical-to-chemical substitution. There is also a financial assessment to whatever extent is possible. However, Harriman noted, "We often do not have that

1. Define Goal
2. ID Chemicals of High Concern
3. Identify Alternatives
4. Prioritize and Pre-Screen Alternatives
5. Alternatives Assessment
   - Technical/Performance Assessment
   - EH&S Assessment
   - Financial Assessment
6. Analyze Information
7. Select Alternative

FIGURE 6-1 Safer alternatives assessment model.
NOTE: EH&S = environmental health and safety, ID = identify.
SOURCE: IC2 Safer Alternatives Assessment, 2011. Reprinted with permission from IC2.

kind of information. That needs to be left up to companies to do, but we provide what we have." Step 6 is to analyze all that information, and Step 7 is to select the alternative.

One of the alternatives included in the assessment was n-propyl bromide, a chemical that would normally have been discarded in the initial screening because of its human health risks. However, because it is not yet regulated by the EPA, some vendors are selling it to dry cleaners as a substitute that can replace perchloroethylene with no more effort than changing the seals on the dry cleaning machines. "We included it here to try to make sure that dry cleaners were informed before they made that substitution," Harriman explained.

One of the most promising alternatives is wet cleaning, which uses computer-controlled equipment and special detergent packages to clean clothes with very little water; afterward the clothes are put into a dryer until they are almost dry, at which point they are removed from the dryer and finished with special finishing equipment. Wet cleaning saves energy, it saves money, it often saves water, it results in better indoor air quality, and its quality of cleaning equals that done with perc. There are now 11 dedicated wet cleaners in Massachusetts, and TURI is encouraging more dry cleaners to switch to wet cleaning with demonstrations and the provision of information.

In a second case study, Harriman spoke about hexavalent chromium. It is used in defense and aerospace applications, mainly in sealants, primers, and conversion coatings. Hex chrome is known to be carcinogenic in humans and a mutagen and developmental toxicant. Long-term inhalation can lead to lung cancer, perforation of the nasal septum, and asthma. TURI got involved in the search for alternatives in part because in 2011 the Defense Federal Acquisition Regulation Supplement called for industry to come up with alternatives and to get those through the approval process.

In one typical use of hex chrome, a conversion coating containing the material is applied to an aluminum substrate; then a sealant containing hex chrome is used to fill gaps and recesses around the fasteners and joints; a primer containing hex chrome is applied on top of the sealant; and a topcoat is applied on top of that. The initial goal would be to replace the primer and sealant with versions that do not contain hex chrome, with the eventual goal being to find an alternative to the hex chrome conversion coating as well, so that the hex chrome can be done away with altogether.

A number of different companies are involved in the alternatives assessment and performance testing. "We could not do it without all those resources," Harriman said. "The companies end up putting in millions of dollars worth of time in terms of fabricating and testing." Lockheed Martin, for example, is carrying out accelerated corrosion testing, while NASA is doing long-term corrosion testing. TURI does the statistical analysis and writes papers reporting the results.

Part of the project involves bringing the different components of the supply chain together so that people can communicate and understand the needs and constraints of others. The supply chain includes the U.S. Department of Defense, the original equipment manufacturers, the component and material suppliers, and the metal finishers. "The metal finishers, which are sort of at the bottom of that supply chain, have a much harder time meeting some of these requirements, and they do not necessarily get a lot of assistance," Harriman said. "Their customer will say, 'I want a hex-chrome-free finish on that,' but they do not necessarily give them all the technical assistance they need to provide that."

Much of the resistance to the process is coming from the metal finishers, she said. "These products have worked well for them for a long time and they are very resistant to change."

Much still remains to be done, even after appropriate alternatives are found. In many cases changes in the military specifications are required. And changing a sealant may require, for example, a change in the sealant remover. It is important to make sure that the new sealant remover is not something toxic.

"In summary," she said, "our objective is to eliminate the hazard, to adopt safer alternatives where they are available, to do alternatives assessments to avoid regrettable substitutions, and to form these collaborations and partnerships with companies so that the supply chains can benefit from that assistance."

CASE STUDY: BULLITT CENTER

The session's last presenter was Joseph David, Sustainability Program Manager at Point32, a real estate company in Seattle, Washington, focusing on land use development and construction. David is in charge of Point32's effort to secure the Living Building Challenge certification for the Bullitt Center, and he spoke of his experiences as a consumer trying to navigate the world of selecting toxic-free materials.

The Bullitt Center, completed in April 2013, is a 52,000-square-foot office building in the Central District area of the Capitol Hill neighborhood of Seattle. It was designed to be green in a number of ways. It generates all of its own electricity on site through a 14,000-square-foot solar array on its roof. It is also designed to be a net zero user of water, with much of its water supplied by capturing the rainwater that falls on its roof in a 56,000-gallon cistern in the basement. The water is filtered and treated on site for potable and nonpotable uses in the building. The goal is also to produce zero net waste. The building uses composting toilets and creates field-ready compost in the basement.

David summarized the goals of the Bullitt Center project with a quote from Denis Hayes, the director of the Bullitt Foundation: "Our desire is to open a wedge into the future so that we can see what is possible in a contemporary office building." The project needs to be profitable, David said, as it is owned by a small philanthropy as part of its investment portfolio, but it also aims to be a game changer. Hayes had also said that if the building turns out to be a unique project in 5 years, then the foundation will have missed its goal entirely. "We want to uncover successes, failures, and share that with the green building community throughout the world," David said.

In building the Bullitt Center, David said, another goal was to minimize the use of toxic materials. One motivation was the presence of a variety of toxic chemicals in Puget Sound, many of which were deposited there because of runoff from the roofs of buildings, the city streets, and other sources. The developers of the center did not want to contribute to that problem.

To that end, the developers decided to adhere to an environmental building standard called the Living Building Challenge, using various issues in green construction such as responsible site selection, water management, energy conservation, and the use of green materials. It was the materials part of the standard that was most unfamiliar, David said. "We talk about green materials in the context of recycled content or where the product is sourced, but the issue of toxicity had not really come up."

To adhere to the Living Building Challenge standard it was necessary to avoid using any materials from a list of 362 prohibited chemicals, including asbestos, chlorofluorocarbons, formaldehyde, halogenated flame retardants, lead, petrochemical fertilizers and pesticides, phthalates, and PVC. "We have to prove to an auditor that, with a whole bunch of documentation, we have done our very best to avoid these chemicals."

To do that it was necessary to check out the material safety data sheet for each potential building material. These data sheets list every component of a material, and these components can be checked against the list of prohibited chemicals. At first it seemed overwhelming, David said. "We are architects, engineers, builders, and contractors. We are not chemists or toxicologists." If they saw formaldehyde on the data sheet, it was easy enough to cross that building material off the list and look for an alternative. But what to do about a chemical like decabromodiphenyl oxide? "We do not know if this is good or bad. What do we do?"

Pretty quickly, he said, they realized that they should not be working with the names of chemicals but rather with the Chemical Abstract Service Registry numbers (CAS numbers). "Using available databases such as the Pharos Project database, we were able to take these CAS numbers and quickly vet them against known red list or avoided chemical lists, including the Living Building Challenge red list."

The vetting process they developed grouped the building materials into three major categories. There were materials, which were things like paint or sealant or adhesives. "It typically comes in a five-gallon pail and we typically use a brush or a gun to install it," he said. There are material safety data sheets available for such materials, which can be tested against the red list.

A second category consists of things they referred to as "articles"— small components such as a ball valve for a plumbing joint. Such articles do not come with a material safety data sheet.

The third category is assemblies, things that have more than 10 parts. An example would be a water pump, which might contain hundreds of distinct components—the wiring, the circuitry, and so on. Again there is no material safety data sheet provided. "That is a pretty daunting task to figure out what red list chemicals might be present in something like that," David said.

Thus, the vetting process was simplest for materials: Get the material safety data sheet, extract the CAS numbers, and run the numbers through the Pharos Project database. However, they discovered that the data sheets generally disclosed only 15 to 20 percent of the chemicals used in the product. Sometimes the materials on the data sheet would be stamped as a trade secret or proprietary. "We essentially began a campaign to cold call the manufacturers of every product in the building and ask for cooperation in confirming that none of our red-listed chemicals were used in their products," David said. That took a huge amount of time and energy for both the project team and the manufacturer.

The process was more difficult for articles. They could ask a manufacturer for a data sheet for each of the materials that made up an article, or they could have the manufacturer sign a letter confirming that none of the red-listed materials were in any of the materials that made up the article.

Assemblies were an even bigger challenge. "It is a really difficult conversation to have with a manufacturer," David said. "It is hard to know where all the distinct parts and pieces come from that make up that pump. . . . Sometimes we were able to get a blanket statement saying, Yes, we can confirm that 10 percent by weight or volume of this product does not contain red list chemicals, but beyond that, it was difficult."

This process was carried out for more than 1,000 products over the 3 years it took to design and construct the Bullitt Center. "We quickly realized we needed to standardize this process," David said, so they developed a building material information request form that asked a manufacturer to answer basic questions about a product: Where is the product manufactured? Where are the source materials coming from? Did they contain the red list chemicals? What is the content of volatile organic compounds? What is the recycled content? And so on. "We sent that out to all the manufacturers. In most cases we got some level of participation and got these forms back and made the best decision we could about which products to use for our building."

From the manufacturer's point of view, it was equally burdensome, as the Bullitt Center was not the only project with such questions. "The manufacturers are receiving these questions from dozens, hundreds, maybe even thousands of projects," he said. "All the forms look slightly different. The questions are very similar . . . but it is a tremendous burden on the manufacturers to field all these questions."

The construction industry is struggling with the issue of how much chemical disclosure is appropriate. At this point, David said, there is a spectrum of disclosure. At one end is the material safety data sheet that virtually every product has. The problem with it is that it is difficult to extract all the information you need to make an informed decision about the constituent chemicals in that product. On the other end of the spectrum is a new label that has recently come into use. The Declare label requires manufacturers to publicly disclose 99 percent of the constituents in their products. That is convenient for consumers, but it presents some serious difficulties related to proprietary formulations and trade secrets. In the middle of the spectrum is a form called the Health Product Declaration. It is a reporting protocol that many manufacturers and consumers have agreed on as a good way to standardize information about chemical constituents. It provides more useful information than the material safety data sheets but offers less of a threat to a business's proprietary information.

In closing, David described a particular episode during the construction of the Bullitt Center. A product called Fastflash from a company called Prosoco was being considered for use in the building's liquid applied air barrier. "This is arguably one of the most important layers in the building," he said. "It keeps the rain out. It keeps the warm conditioned air in." They got the material safety data sheet for Fastflash and discovered there were a number of proprietary chemicals, so they called up Prosoco and asked if any red list chemicals were in it. Yes, it contained a type of phthalate which is what allowed it to stretch and flex. "I said thanks for working with us. We cannot use your product on this building."

One week later, however, the company called back to say that their engineers had been working on a reformulation of the product that could get rid of the phthalates. If they could have 6 months for research and development they should be able to help. Then, 5 months after that, the company called to say that it had succeeded. "Come on down to our lab. We just ran this reformulated product through 500 hours of hurricane testing in our test chamber. It is performing quite well." And that material is what is now installed in the Bullitt Center—a reformulated, phthalate-free version of the original material.

Seeing the success of that product, the manufacturer decided on a wholesale elimination of phthalates from its entire product line, and now all its products are phthalate-free and compliant with the Living Building Challenge red list, David said. "I think this is really a testament to the benefit of entering dialogue between consumer and manufacturer."

DISCUSSION

In the discussion following the presentation, the first question concerned the attitudes toward green chemistry that are generally found within companies, laboratories, and other organizations. Trisha Castranio of NIEHS said that while laboratory employees generally do not want to have to worry about green programs, they do not want to be around toxic materials either, so they are willing to participate in programs to reduce them. Nonetheless, she said, the major push for green materials will need to come from consumers. They will be the ones driving the movement toward for green chemistry.

Zephanie Jordan of Johnson & Johnson said that attitudes within her company differ from department to department. For example, public affairs people believe it is important to the company's leadership position that it continue to move forward on sustainability, but those in the supply chain and research and development are very sensitive to the disruptions that the move to green chemistry can cause. It is a real tension, she said.

Connie Deford of Dow echoed Jordan's observation. Many new graduates coming into the company are drawn by its sustainability programs and are very supportive of the move to green chemistry, while the employees who have been there for a while recognize how very challenging and expensive it can be for the company to make the major changes in the materials that they produce and use.

David Constable of the ACS Green Chemistry Institute observed that most companies have people who are working to institute green chemistry, but there are many competing demands, and many people worry that the move to green chemistry may affect the quality of their products or something else in a negative way.

Beverley Thorpe of Clean Production Action commented that one of the major obstacles to the movement to green chemistry is disclosure. Without disclosure, it is impossible to make informed decisions, but many companies resist making such disclosures out of fears of losing a competitive advantage.

A follow-up question concerned how best to share the necessary information about products while still protecting necessary confidentiality. Jordan answered that her company, Johnson & Johnson, struggles with that issue, particularly as it relates to how much information it provides to consumers. One approach, she suggested, is to provide the information to regulators. There are also some voluntary systems that Johnson & Johnson participates in and to which the company provides information about its formulations.

Deford of Dow agreed that it is a challenging issue. One approach is to use nondisclosure agreements. Also, she said, another option is to provide information on the product family rather than the specific chemical as the health and environmental profiles are very similar, or another option is to provide the requested information to a third party.

Constable suggested that third-party certification can solve some of the problem. Having a material being certified as not containing any of a list of chemicals can be enough for certain customers.

Liz Harriman suggested that the confidentiality issues may not be as serious as some have suggested because of the increasing ability to analyze products and determine exactly what materials they contain. "I am not sure that there is really as much proprietary information as business would like to think there is," she said.

Joseph David of Point32 said that while some consumers are pushing for full disclosure of materials in order to make informed assessments, having a third-party verifier do independent assessments of products is also workable. Either model works pretty well, he said.

Castranio pointed out, however, that there are a large number of different groups offering different sorts of seals of approval, and it is very difficult for consumers to learn enough about them to know which to trust. David agreed and suggested it would be useful to get industry agreement on a gold standard in certification in each sector.

An audience member asked how information is collected about chemicals from other countries. David said that in building the Bullitt Center almost all of the products were produced within 1,000 kilometers of Seattle and there were very few instances where they had to go to an international source for a material in the building. In those cases there was a partner in the United States that was able to convey the questions and get information and whatever disclosure material was available from the suppliers.

Constable said that there are not really any good mechanisms to get good information about what is in the chemical formulations that come from places like China or India. Thorpe suggested that, given the growing push for green products, the difficulty of proving that materials from other countries do not contain any hazardous chemicals might lead to a certain amount of relocation of the supply chain back to the United States.

REFERENCE

IC2 Safer Alternatives Assessments. 2011. *IC2 safer alternatives assessments.* Available at http://www.ic2saferalternatives.org (accessed January 28, 2014).

7

Reflections on the Workshop and Concluding Remarks

In the workshop's final session, various speakers synthesized, commented on, and expanded on the presentations and discussions that had taken place on the previous day and a half.

Kimberly Thigpen Tart, Program Analyst in the Office of Policy, Planning, and Evaluation at the National Institute of Environmental Health Sciences, commented on topics raised during the presentations looking at the challenge of chemicals in today's society (see Chapter 2). She noted that Lynn Goldman's remarks on the Toxic Substances Control Act (TSCA) of 1976[1] and the difficulty of determining how many chemicals are in use in the United States (and in what forms) helped set the stage for later presentations. From a public health perspective, William E. Halperin described the difficulty in approaching industrial chemical assessments when at least five different paradigms could be used: industrial hygiene, prevention, surveillance, embeddedness, and dose response. Thigpen Tart highlighted that a better public health approach, as articulated by Halperin, may try to focus assessment and management efforts on reducing chemical exposures to everyone in order to reach a greater number of people affected by significant exposures, not just those with the highest level of exposures. Efforts are under way, including the National Conversation on Public Health and Chemical Exposures, to join diverse public and private stakeholders in working to ensure that chemicals are used and managed in ways that protect the health and safety of individuals. Thigpen Tart noted that some workshop participants expressed concerns that the public's voice in chemical risk evaluation is being lost, and that the resource base for addressing community concerns has been "tattered" at

---

[1] Toxic Substances Control Act of 1976, Public Law 94-469, 94th Congress.

every level. She highlighted remarks from Goldman that social media and related technologies show promise for empowering people to connect and network outside traditional structures, but that public health officials and researchers should be involved in research translation to ensure that accurate and appropriate information is disseminated at the community level, and that people and communities know how best to apply this information to their decision-making processes.

Bernard Goldstein, Emeritus Dean and Emeritus Professor of Environmental and Occupational Health at the University of Pittsburgh Graduate School of Public Health, discussed some of the common themes emerging from the discussions on current programs for safeguarding the public from the potential health risks of industrial chemicals. After the presentation on the Registration, Evaluation, Authorisation, and Restriction of Chemicals (REACH) legislation, by Canice Nolan, the talks by the other presenters could be summarized in two words, he said: Fix TSCA. "All three—Wendy Cleland-Hamnett from the U.S. Environmental Protection Agency (EPA), Richard Denison from Environmental Defense Fund, and Michael Walls from the American Chemistry Council—agreed on the need to move something forward in legislation," he said. "I found them pretty close in terms of the details."

Beyond that, Goldstein addressed what he described as "hidden assumptions" in the field. One lies in the area of toxicology testing. There are about 20,000 new chemicals that have come to market since TSCA was passed. If you assume that existing toxicology is good enough to pick up, say, 99 percent of the chemicals that are causing reproductive and developmental effects, that still leaves 1 percent, or 200 chemicals that are potential reproductive and developmental toxins that the EPA has allowed into commerce. "I do not think we are 99 percent effective," he said. "I think we have a long way to go. I picked reproductive development. You can pick neurotoxins. You can pick chronic disease issues. You can pick lots of the other systems for which our test are inadequate to predict the effects of chemicals."

The assumption seems to be that the problem can be fixed by doing more toxicology, Goldstein said. "I think a lot of the presentations in the fourth panel assumed that as long as we did the toxicology testing that we know how to do, we would be fine. No. We will not be fine." A key issue, he said, will be "how do you develop the Toxic Substances Control Act in a way that makes sure that we have the best testing rather than just saying you have to test everything."

A second issue is how best to promote innovation by industry to decrease pollution emissions. Do thresholds exist to promote innovation in approaches to meet the threshold, or the opposite? Will the size of the threshold make a difference? Will the kind of testing that is required make a difference? Will putting chemicals in different levels help or hurt innovation? Goldstein mentioned an experience in which an emission threshold put in under the Clean Air Act led a company to replace a solvent with another solvent that had one-tenth the volatility so it would not exceed the reporting levels, but one-hundredth the odor threshold so that now the community smelled the emissions. Was this new solvent more toxic? "I am not sure whether it was or not, but it is an example of how we sometimes drive toward agents that may be more toxic." Thus, a key question is, Are we able to design TSCA in a way that allows for innovation and does not drive us toward chemicals that may be more of a problem?

Finally, Goldstein said, transparency is a difficult issue to deal with. There is certain information that businesses need to keep confidential in order to maintain their intellectual property and competitive advantage, he said, but industry has shown over and over again that it uses the excuse of confidential business information to hide things that it does not need to hide—and sometimes things that it should not hide. How does one craft a policy that protects innovation by allowing companies to keep some things secret while maintaining the distinction between things that need to be kept secret and those that do not?

Hal Zenick, Director of the National Health and Environmental Effects Research Laboratory at the EPA, also provided some observations concerning the presentations, particularly those from the session on improved approaches to chemical prioritization (see Chapter 5). One thing that struck him forcibly, he said, is that the discussions were far from having a one-size-fits-all approach. The prioritizations were aimed at different groups—some at populations, others at communities, for example—and were also accountable to different sets of stakeholders. "I think when we begin to look at a system for prioritization, it probably is not going to be widely applicable across these different venues and populations," he said. Although all the prioritization approaches more or less use a matrix that characterizes hazard on one axis and exposure on the other axis, Zenick noted that the level of precaution tends to differ across the approaches with some groups moving ahead boldly and others conservatively.

Zenick felt that biomonitoring was mentioned repeatedly and he explained that "the opportunity for acquiring biomonitoring data is becoming greatly enhanced." The example of the state laboratory capabilities brought up in the discussion time of that session could be a resource for investigating biomonitoring and environmental monitoring data. Zenick stated that those resources could be used to provide exposure information in geographic units more valuable than what can be extracted at the national level.

Lynn Goldman offered a few common themes that appeared from all of the case studies presented in Session 5. First, she said, it is clear that there are many drivers for reducing the risk of chemicals; some of them have to do with reducing costs, some with reducing risk, and some with such intangibles as the reputation of products and companies. Second, everyone seemed to believe that a life-cycle approach was very important. "If you are just working at the end of pipe, you can miss things that are important."

Third, information is both a driver and an enabler. That seemed to be a very important point, Goldman said, as well as the fact that our lack of information about so many chemicals leads people to rely on lists that are created on the basis of very limited existing information and assessments that have been done. She noted that we look under the same lamppost where the light has shone for years, and at the same substances that have been tested again and again, and measured again and again, while there are so many other substances about which we know so little that we cannot factor them into assessments.

A fourth common point was the importance of educating or working with the organic and analytic chemists themselves whose work lies at the core of the decisions that will ultimately be made. "The decisions that they make about how methods should be done can have enormous implications," Goldman said. If an analytic method required by the government uses a large amount of methylene chloride, then that is going to drive the use of methylene chloride in laboratories across not only the United States, but also other countries.

Frank Loy, chair of the Roundtable, then commented on the workshop as a whole. He highlighted the interesting, thoughtful work done on the prioritization of chemicals for risk assessment in the state of California and in Canada. "I thought this work was quite impressive and useful," he said. He noted the differing views that were expressed related to the European Union's REACH program. In his opinion, the REACH program has gone over well with relative ease, and while Europe is not the United States, there are lessons to learn from this program when updating TSCA. One missing element from the workshop was someone representing labor and workers who may be exposed to toxic chemicals. "We need to recognize that some people are more exposed to these chemicals than others. If there are categories of people like that, we need to have them participate in these sessions," he said. Loy stated that all the workshop presentations and discussions were thoughtful, and he hopes the views and opinions expressed will help inform next steps.

A

Glossary[1]

Adverse outcome pathway: a conceptual framework that portrays existing knowledge concerning the linkage between a direct molecular initiating event and an adverse outcome, at a level of biological organization relevant to risk assessment (EPA, 2013a).

Alternative chemicals: chemicals within the same functional-use group across a consistent and comprehensive set of hazard end points that exhibit safer health and environmental profiles than chemicals of potential concern (EPA, 2013b).

Bioaccumulation: general term describing a process by which chemicals are taken up by an organism either directly from exposure to a contaminated medium or by consumption of food containing the chemical (EPA, 2012b).

Biological pathway altering dose: exposure level at the low end of the distribution of the pathway-altering dose that is calculated with standard risk assessment approaches and in vitro assays to quantitatively characterize the chemical with high-throughput methods and estimate the external dose that would be required to perturb a biological pathway (Judson et al., 2011).

Certain internationally classified substances: are among those identified as priorities for action for the second phase of the Chemicals Management Plan in Canada. The selection of these internationally classified substances for action is based on the categorization process completed in 2006, and new information received as part of the first phase of the Chemicals Management Plan (Government of Canada, 2012a).

---

[1] Most definitions in the glossary are direct quotations from the cited material.

Chemical Abstract Service Registry numbers (CAS numbers): a unique identifier assigned by the Chemical Abstract Service of the American Chemical Society to identify a chemical substance or molecular structure when there are many possible systematic, generic, proprietary, or trivial names. CAS numbers are used in public and private databases and on Material Data Safety Data Sheets (CAS, 2014).

Chemical and Product Categories (CPCat): a database of information on how chemicals are used. The data are divided into types of uses and specific products the chemicals are used in (EPA, 2014b).

Chemicals Management Plan: launched in 2006, the government of Canada uses the Chemicals Management Plan to set clear priorities for the assessment and management of chemical substances (Government of Canada, 2014).

Clinical evaluation: an evaluation of the safety of specific concentrations and mixtures of ingredients in consumer products; scientists and clinicians look for adverse reactions, mixture interactions, and other safety considerations (Johnson & Johnson, 2014a).

Co-exposures: contact or occurrence of more than one exposure (EPA, 2012a).

Concentration at steady state (Css): the concentration of a drug or chemical in a body fluid—usually plasma—to achieve a steady state where rates of drug administration and drug elimination are equal (Boston University School of Medicine, 2004).

Direct and indirect exposure: direct exposure involves physical contact made between the chemical agent and individual (e.g., skin, lungs, gut), whereas indirect exposure involves transport of the chemical from the source to the environment to the individual (e.g., consumption of fruits and vegetables with pesticide residues) (EPA, 2012a).

Dose: total amount of a substance administered to, taken up, or absorbed by an organism, organ, or tissue (IUPAC, 2007).

Dose–response relationship: association between dose and the incidence of a defined biological effect in an exposed population usually expressed as a percentage (IUPAC, 2007).

Eco exposome: the extension of exposure science from the point of contact between stressor and receptor inward into the organism and outward to the general environment, including the ecosphere (NRC, 2012).

Embeddedness: a state of being located or secured within a larger entity or context (Mayhew, 2009).

Endocrine Disruptor Screening Program: the 1996 amendments to the Safe Drinking Water Act authorized the U.S. Environmental Protection Agency (EPA) to screen substances that may be found in sources of drinking water for endocrine disruption potential. The EPA established the Endocrine Disruptor Screening Program, a scientific advisory committee, to advise the EPA on establishing a program to carry out Congress's directives (EPA, 2011).

EPA: the U.S. Environmental Protection Agency; mission to protect human health and the environment (EPA, 2014g).

ExpoCast: the EPA evaluates the potential risks of the manufacture and use of thousands of chemicals. To assist with this evaluation, the EPA scientists developed a rapid, automated (high-throughput) model using off-the-shelf technology that predicts exposures for thousands of chemicals. These predictions are being used to prioritize the order in which chemicals should be evaluated further. The EPA refers to this research effort as ExpoCast (EPA, 2014d).

Exposure: the concentration, amount, or intensity of a particular physical or chemical agent or environmental agent that reaches the target population, organism, organ, tissue, or cell, usually expressed in numerical terms of concentration, duration, and frequency. Process by which a substance becomes available for absorption by the target population, organism, organ, tissue, or cell, by any route (IUPAC, 2007).

Global Aquatic Ingredient Assessment (GAIA): a tool from Johnson & Johnson used to assist its formulators with selecting ingredients that have reduced environmental impacts at the end of use phase (Johnson & Johnson, 2014b).

Globally Harmonized System of Classification and Labeling Chemicals (GHS): a system for addressing the classification of chemicals by types of hazard and proposing harmonized hazard communication elements, including labels and safety data sheets. It aims at ensuring that information on physical hazards and toxicity from chemicals be available in order to enhance the protection of human health and the environment during the handling, transport and use of these chemicals (UNECE, 2014).

Grandfathered: a grandfather clause is a part of a law that says the law does not apply to certain people and things because of conditions that existed before the law was passed (Merriam-Webster, 2014).

Green chemistry: the design of chemical products and processes that reduce or eliminate the generation of hazardous substances (EPA, 2014e).

GreenScreen® for Safer Chemicals: a method for comparative chemical hazard assessment. It is used by a wide range of professionals, government bodies, nonprofits, businesses, formulators, and product developers, and anybody interested in assessing the inherent hazards of chemicals and their potential effect on human health and the environment (Clean Production Action, 2014).

Hazard: set of inherent properties of a substance, mixture of substances, or a process involving substances that, under production, usage, or disposal conditions, make it capable of causing adverse effects to organisms or the environment, depending on the degree of exposure; in other words, it is a source of danger (IUPAC, 2007).

High-content data: results from data-rich techniques, such as genomics, proteomics, and high-throughput screens (NRC, 2010).

High Production Volume Challenge Program: the EPA program where companies are challenged to make health and environmental effects data publicly available on chemicals produced or imported in the United States in quantities of 1 million pounds or more per year (EPA, 2013c).

High-throughput screening assays: in vitro biochemical- and cell-based assays and nonrodent animal models for toxicology testing that allow for much higher throughput at a much reduced cost. In some assays, many thousands of chemicals can be tested simultaneously in days (National Toxicology Program, 2014).

In use testing: a review and evaluation of how volunteers use a product in their homes before the product can go to market (Johnson & Johnson, 2014a).

In vitro potency: expression of relative toxicity of an agent involving isolated organ, tissue, cell, or biochemical systems as compared to a given or implied standard or reference (IUPAC, 2007).

Intrinsic clearance rate: volume of plasma or blood from which a substance is completely removed in a period of time under unstressed conditions (IUPAC, 2007).

Least burdensome: the Toxic Substances Control Act (TSCA) of 1976 indicates that the EPA should apply the least burdensome means of adequately protecting against the unreasonable risk. In developing a rule, TSCA directs the EPA to consider and publish a statement with respect to (1) the effect of the chemical substance being regulated on health and the magnitude of exposure of humans to the substance, (2) the effects of such substance on the environment and the magnitude of exposure of the environment to the substance, (3) the benefits of such substance for various uses and the availability of substitutes for such uses, and (4) the reasonably ascertainable economic consequences of the rule, after consideration of the effect on the national economy, small business, technological innovation, the environment, and public health (EPA, 2014j).

Level of precaution: decisions employed to achieve a chosen level of health and environmental protection under conditions of uncertainty (WHO, 2004).

Life-cycle assessment: a systems-based approach to quantifying the human health and environmental impacts associated with a product's life from "cradle to grave." A full life-cycle assessment addresses all stages of the product life cycle and should take into account alternative uses as well as associated waste streams, raw material extraction, material transport and processing, product manufacturing, distribution and use, repair and maintenance, and wastes or emissions associated with a product, process, or service as well as end-of-life disposal, reuse, or recycling (EPA, 2013d).

Material safety data sheet: compilation of information required under the Occupational Safety and Health Administration Communication Standard on the identity of hazardous substances, health and physical hazards, exposure limits, and precautions (IUPAC, 2007).

Median lethal dose (LD$_{50}$): statistically derived median dose of a chemical or physical agent expected to kill 50 percent of organisms in a given population under a defined set of conditions (IUPAC, 2007).

Molecular epidemiology: the use of the techniques of molecular biology in the study of the distribution and determinants of disease occurrence in human populations (Foxman and Riley, 2001).

NexGen: Advancing the Next Generation of Risk Assessment (NexGen) is a component of the EPA's Chemical Safety for Sustainability Research program and is focused on fostering practical applications on new methods in risk assessment (EPA, 2013e).

Pattern: information or data on human activity used in exposure assessments (EPA, 2012a).

Persistence: attribute of a substance that describes the length of time that the substance remains in a particular environment before it is physically removed or chemically or biologically transformed (IUPAC, 2007).

Pharos Project: a database that contains product and hazard information from more than 300 companies to help users locate materials to meet their current needs; a platform where users can discuss what products support environmental, health, and social equity practices (Pharos, 2014).

Prioritization: organization of chemical substances in order of priority so the most important receives detailed evaluation and assessment. Many approaches consider the degree of hazard and extent of exposure potential when prioritizing chemicals (ACC, 2011).

Production volume: volume of chemicals produced.

REACH: the European Commission Regulation 1907/2006 on the Registration, Evaluation, Authorisation, and Restriction of Chemicals. REACH streamlines and improves the former legislative framework on chemicals of the European Union, and makes industry responsible for assessing and managing the risks posed by chemicals and providing appropriate safety information to their users (European Commission, 2013).

Reverse engineering: the separation, identification, and quantitation of ingredients in a formulation (Chemir, 2014).

Risk: probability of adverse effects caused under specified circumstances by an agent in an organism, a population, or an ecological system; probability of a hazard causing an adverse effect; and expected frequency of occurrence of a harmful event arising from such an exposure (IUPAC, 2007).

Risk assessment: identification and quantification of the risk resulting from a specific use or occurrence of a chemical or physical agent, taking into account possible harmful effects on individuals or populations exposed to the agent in the amount and manner proposed and all the possible routes of exposure (IUPAC, 2007).

SIN (Substitute It Now!) List: a nongovernmental organization driven project (led by ChemSec) to assist with the REACH legislative process and give guidance to companies on safer chemical substitutions for potentially hazardous chemicals (ChemSec, 2013).

SNAcs: as part of the Chemical Management Plan, Environment Canada or Health Canada may place restrictions on reintroduction and new uses of existing chemical substances using Significant New Activity (SNAc) provisions. The SNAc provisions are very similar to SNURs issued by the EPA (Government of Canada, 2012b).

SNUR: after issuing a Consent Order, the EPA generally promulgates a significant new use rule (SNUR) that binds all manufacturers and processors to the terms and conditions contained in the Consent Order. The SNUR requires that manufacturers (which include importers) and processors of certain substances notify the EPA at least 90 days before beginning any activity that the EPA has designated as a "significant new use." This allows the EPA the opportunity to review and if necessary prevent or limit potentially adverse exposure to, or effects from, the new use of the substance (EPA, 2014f).

Sourcing raw materials: the process of obtaining raw materials from suppliers (Johnson & Johnson, 2014a).

Substance Groupings Initiative: plan to assess and manage, where appropriate, the potential health and ecological risks associated with nine groupings of chemical substances (identified based on structural or functional similarities) selected for further action based on a categorization exercise completed as part of the Chemicals Management Plan (defined earlier) (Government of Canada, 2013).

Substitution: the move from problematic chemicals to safer chemicals (alternative chemicals), while minimizing the likelihood of unintended consequences (EPA, 2014c).

Surveillance: systematic ongoing collection, collation, and analysis of data and the timely dissemination of information to those who need to know in order that action can be taken to initiate investigative or control measures (IUPAC, 2007).

Sustainability: creates and maintains the conditions under which humans and nature can exist in productive harmony, that permit fulfilling the social, economic, and environmental requirements of present and future generations (EPA, 2014k).

Tox21: an interagency federal research program to use robotics technology to screen thousands of chemicals for potential toxicity, using screening data to predict the potential toxicity of chemicals and developing a cost-effective approach for prioritizing the thousands of chemicals that need toxicity testing (EPA, 2014a).

ToxCast™: a multiyear effort launched in 2007 that uses automated chemical screening technologies (high-throughput screening assays) to expose living cells or isolated proteins to chemicals. The cells or proteins are then screened for changes in biological activity that may suggest potential toxic effects and eventually potential adverse health effects (EPA, 2014i).

Toxicity: capacity to cause injury to a living organism defined with reference to the quantity of substance administered or absorbed, the way the substance is administered and distributed in time (single or repeated doses), the type of severity of injury, the time needed to produce the injury, the nature of the organism(s) affected, and other relevant conditions (IUPAC, 2007).

Toxicity assessment: the purpose of the toxicity assessment is to weigh available evidence regarding the potential for particular contaminants to cause adverse effects in exposed individuals and to provide, where possible, an estimate of the relationship between the extent of exposure to a contaminant and the increased likelihood and/or severity of adverse effects (EPA, 1989).

Toxics Use Reduction Institute (TURI): established by the Massachusetts Toxics Use Reduction Act of 1989, TURI collaborates with businesses, community organizations, and government agencies to reduce the use of toxic chemicals, protect public health and the environment, and increase competitiveness of Massachusetts businesses (TURI, 2014).

TOXLINE: Toxicology Literature Online; data network of references from toxicology (NLM, 2014).

TSCA: the Toxic Substances Control Act (TSCA) of 1976 provides the EPA with authority to require reporting, record-keeping and testing requirements, and restrictions relating to chemical substances and/or mixtures. Certain substances are generally excluded from TSCA, including, among others, food, drugs, cosmetics, and pesticides (EPA, 2014h).

Virtual tissue model: innovative paradigms for understanding disease progression in silico cross-scale models of cellular organization and emergent functions. Tissues are the clinically relevant level for diagnosing and treating the transition from normal to adverse states in chemical-induced toxicities leading to cancer, immune dysfunction, developmental defects, and more (EPA, 2013f).

REFERENCES

ACC (American Chemistry Council). 2011. *ACC prioritization screening approach.* Available at http://www.americanchemistry.com/Prioritization-Document (accessed April 4, 2014).

Boston University School of Medicine. 2004. *Glossary of terms and symbols used in pharmacology.* Available at http://www.bumc.bu.edu/busm-pm/academics/resources/glossary/#c (accessed April 4, 2014).

CAS (Chemical Abstracts Service). 2014. *CAS: A division of the American Chemical Society. FAQs.* Available at https://www.cas.org/about-cas/faqs (accessed April 4, 2014).

Chemir. 2014. *Definitive deformulation.* Available at http://www.chemir.com/deformulation.html (accessed April 4, 2014).

ChemSec. 2013. *What we do: SIN List.* Available at http://www.chemsec.org/what-we-do/sin-list (accessed April 4, 2014).

Clean Production Action. 2014. *GreenScreen® for Safer Chemicals.* Available at http://www.greenscreenchemicals.org (accessed April 4, 2014).

EPA (U.S. Environmental Protection Agency). 1989. Chapter 7: Toxicity assessment. In *Risk assessment guidance for Superfund Volume I human health evaluation manual (Part A).* Washington, DC: EPA. Available at http://www.epa.gov/oswer/riskassessment/ragsa/pdf/rags_a.pdf (accessed April 4, 2014).

EPA. 2011. *Endocrine Disruptor Screening Program (EDSP): EDSP background.* Available at http://www.epa.gov/scipoly/oscpendo/pubs/edsp overview/background.htm (accessed April 4, 2014).

EPA. 2012a. *Thesaurus of terms used in microbial risk assessment—5.5 exposure.* Available at http://water.epa.gov/scitech/swguidance/standards/criteria/health/microbial/t55.cfm#T55030 (accessed April 4, 2014).

EPA. 2012b. *Waste and cleanup risk assessment glossary.* Available at http://www.epa.gov/oswer/riskassessment/glossary.htm#c (accessed April 4, 2014).

EPA. 2013a. *Adverse outcome pathway (AOP) wiki.* Available at http://www.epa.gov/research/priorities/docs/aop-wiki.pdf (accessed April 4, 2014).

EPA. 2013b. *Chemical alternatives assessment.* Available at http://www.epa.gov/sustainability/analytics/chem-alt.htm (accessed April 4, 2014).

EPA. 2013c. *High Production Volume (HPV) challenge.* Available at http://www.epa.gov/hpv (accessed April 4, 2014).

EPA. 2013d. *Life-cycle assessment.* Available at http://www.epa.gov/sustainability/analytics/life-cycle.htm (accessed April 4, 2014).

EPA. 2013e. *NexGen: Advancing the next generation of risk assessment.* Available at http://www.epa.gov/risk/nexgen (accessed April 4, 2014).

EPA. 2013f. *What are virtual tissues?* Available at http://www.epa.gov/ncct/virtual_tissues/what.html (accessed April 4, 2014).

EPA. 2014a. *Computational toxicology research: Tox21.* Available at http://epa.gov/ncct/Tox21 (accessed April 4, 2014).

EPA. 2014b. *CPCat: Chemical and product categories.* Available at http://actor.epa.gov/cpcat/faces/basicInfo.xhtml (accessed April 4, 2014).

EPA. 2014c. *Design for the environment: Alternatives assessment.* Available at http://www.epa.gov/dfe/alternative_assessments.html (accessed April 4, 2014).

EPA. 2014d. *ExpoCast^{TM}.* Available at http://www.epa.gov/ncct/expocast (accessed April 4, 2014).

EPA. 2014e. *Green chemistry.* Available at http://www2.epa.gov/green-chemistry (accessed April 4, 2014).

EPA. 2014f. *New chemical consent orders and significant new use rules (SNURs).* Available at http://www.epa.gov/oppt/newchems/pubs/cnosnurs.htm (accessed April 4, 2014).

EPA. 2014g. *Our mission and what we do.* Available at http://www2.epa.gov/aboutepa/our-mission-and-what-we-do (accessed April 4, 2014).

EPA. 2014h. *Summary of the Toxic Substances Control Act.* Available at http://www2.epa.gov/laws-regulations/summary-toxic-substances-control-act (accessed April 4, 2014).

EPA. 2014i. *ToxCast^{TM}: Advancing the next generation of chemical safety evaluation.* Available at http://www.epa.gov/ncct/toxcast (accessed April 4, 2014).

EPA. 2014j. *TSCA Section 6 actions.* Available at http://www.epa.gov/oppt/existingchemicals/pubs/sect6.html (accessed April 4, 2014).

EPA. 2014k. *What is sustainability?* Available at http://www.epa.gov/sustainability/basicinfo.htm (accessed April 4, 2014).

European Commission. 2013. *REACH: Registration, evaluation, authorization, and restriction of chemicals.* Available at http://ec.europa.eu/enterprise/sectors/chemicals/reach/index_en.htm (accessed April 4, 2014).

Foxman, B., and L. Riley. 2001. Molecular epidemiology: Focus on infection. *American Journal of Epidemiology* 153(12):1135–1141.

Government of Canada. 2012a. *Internationally classified substance groups.* Available at http://www.chemicalsubstanceschimiques.gc.ca/group/internat/index-eng.php (accessed April 4, 2014).

Government of Canada. 2012b. *The Significant New Activity (SNAc) approach.* Available at http://www.chemicalsubstanceschimiques.gc.ca/plan/approach-approche/snac-nac-eng.php (accessed April 4, 2014).

Government of Canada. 2013. *The Substance Groupings Initiative.* Available at http://www.chemicalsubstanceschimiques.gc.ca/group/index-eng.php (accessed April 4, 2014).

Government of Canada. 2014. *Chemicals Management Plan.* Available at http://www.chemicalsubstanceschimiques.gc.ca/plan/index-eng.php (accessed April 4, 2014).

IUPAC (International Union of Pure and Applied Chemistry). 2007. Glossary of terms used in *Toxicology, 2nd edition. Pure and Applied Chemistry* 79(7):1153–1344. Available at http://sis.nlm.nih.gov/enviro/iupacglossary/glossarym.html (accessed April 2, 2014).

Johnson & Johnson. 2014a. *Our five-level safety assurance process.* Available at http://www.safetyandcarecommitment.com/safety-promise (accessed April 4, 2014).

Johnson & Johnson. 2014b. *Product disposal: End of use.* Available at https://www.jnj.com/caring/citizenship-sustainability/strategic-framework/end-of-life (accessed April 4, 2014).

Judson, R. S., R. J. Kavlock, R. W. Setzer, E. A. Hubal, M. T. Martin, T. B. Knudsen, K. A. Houck, R. S. Thomas, B. A. Wetmore, and D. J. Dix. 2011. Estimating toxicity-related biological pathway altering doses for high-throughput chemical risk assessment. *Chemical Research in Toxicology* 24(4):451–462.

Mayhew, S. 2009. *A dictionary of geography*. Oxford: Oxford University Press.

Merriam-Webster. 2014. *Grandfather clause*. Available at http://www.merriam-webster.com/dictionary/grandfather%20clause (accessed April 4, 2014).

National Toxicology Program. 2014. *Tox21*. Available at http://ntp.niehs.nih.gov/?objectid=06002ADB-F1F6-975E-73B25B4E3F2A41CB (accessed April 4, 2014).

NLM (U.S. National Library of Medicine). 2014. *Toxicology Literature Online (TOXLINE)*. Available at http://toxnet.nlm.nih.gov/html/TOXLINE.htm (accessed April 4, 2014).

NRC (National Research Council). 2010. *Toxicity pathway-based risk assessment: Preparing for paradigm change: A symposium summary*. Washington, DC: The National Academies Press.

NRC. 2012. *Exposure science in the 21st century: A vision and a strategy*. Washington, DC: The National Academies Press.

Pharos. 2014. *About Pharos*. Available at http://www.pharosproject.net/about/index (accessed April 4, 2014).

TURI (Toxics Use Reduction Institute). 2014. *About: Who we are*. Available at http://www.turi.org/About (accessed April 4, 2014).

UNECE (United Nations Economic Commission for Europe). 2014. *Globally Harmonized System of Classification and Labeling of Chemicals (GHS)*. Available at http://www.unece.org/trans/danger/publi/ghs/ghs_welcome_e.html (accessed April 4, 2014).

WHO (World Health Organization). 2004. *The precautionary principle: Protecting public health, the environment, and the future of our children*. Copenhagen: WHO. Available at http://www.euro.who.int/__data/assets/pdf_file/0003/91173/E83079.pdf (accessed April 4, 2014).

B

Agenda

November 7–8, 2013
Room 100
Keck Center of the National Academy of Sciences
500 Fifth Street, NW, Washington, DC

November 7, 2013

9:00 a.m.   **Welcome**

                Frank Loy, LL.B.
                Roundtable Chair

Session 1: Background and Framing

Objectives:

- Provide an overview of chemical exposures from pesticides, to cosmetics, to food, and to industrial processes and explain why the workshop will focus on new and existing industrial chemicals.
- Briefly highlight what is known now that was not known 20 years ago about the links between chemical hazards and human health.
- Describe a public health approach to industrial chemical assessments based on identifying sentinel health events, delineating a cascade of preventive interventions (primary, secondary, and tertiary), and using public health surveillance to monitor and improve the system.

9:05 a.m.   **Overview of Our Daily Exposure to Chemicals and the Need to Discuss Industrial Chemical Assessments**

                Lynn Goldman, M.D., M.P.H.
                Roundtable Vice-Chair

	Dean, School of Public Health and Health Services George Washington University
9:20 a.m.	**Public Health Approach to Industrial Chemical Assessments**
	William E. Halperin, Dr.P.H., M.D., M.P.H. Chair and Professor Department of Preventive Medicine and Community Health New Jersey Medical School Professor and Associate Dean Rutgers School of Public Health
9:35 a.m.	**National Conversation on Public Health and Chemical Exposures Action Agenda**
	Nsedu Obot Witherspoon, M.P.H. Co-Chair, National Conversation Leadership Council Executive Director Children's Environmental Health Network
9:50 a.m.	**Discussion**
10:20 a.m.	**Break** (15 minutes)

**Session 2: Current Programs for Safeguarding the Public from
Potential Health Risks of Industrial Chemicals:
Successes and Areas for Improvement**

Objectives:

- Provide an overview of what is working and where gaps may be present in a variety of existing regulations and industrial chemical safety programs.
- Explain that the manufacturer of a chemical is primarily informed about its toxicity information but may not know specifically how the chemical will be used, whereas the processer of the chemical is primarily informed about its uses and exposure information (among workers and consumers) and may use the chemical in ways the manufacturer never considered.
- Propose suggestions for how to improve the effectiveness and efficiency of regulatory programs.

Moderator: Dennis Devlin, Ph.D., Environmental Health Advisor, ExxonMobil Corporation

10:35 a.m.	**Global Perspective**

Canice Nolan, Ph.D.
Senior Coordinator for Global Health
European Commission Directorate General for Health and Consumers

10:50 a.m.	**U.S. Federal Perspective**

Wendy Cleland-Hamnett, J.D.
Director, Office of Pollution Prevention and Toxics
Office of Chemical Safety and Pollution Prevention
U.S. Environmental Protection Agency

11:05 a.m.	**NGO Perspective**

Richard Denison, Ph.D.
Senior Scientist
Environmental Defense Fund

11:20 a.m.	**Industry Perspective**

Michael P. Walls, J.D.
Vice President of Regulatory and Technical Affairs
American Chemistry Council

11:35 a.m.	**Discussion**
12:20 p.m.	**Lunch Break** (60 minutes)

Session 3: Models for Environmental Risk Assessment and Exposure Science

Objectives:

- Provide brief summaries of key meetings and reports that have been held and released on environmental risk assessment and exposure science.
- Comment on where all this research may be leading:
  - How will this inform work to improve policy, assessment, and action strategies related to industrial chemical safety programs?

- o What more needs to be done to allow for better modeling, monitoring, and measurement of industrial chemical exposures?
- Present a status update of a new National Research Council committee tasked to develop a decision framework for evaluating potentially safer chemical substitutions.

Moderator: Susan Santos, Ph.D., M.S., Assistant Professor, Rutgers School of Public Health

1:20 p.m.	**Science and Decisions: Advancing Risk Assessment** (report released in 2009) John M. Balbus, M.D., M.P.H. Senior Advisor for Public Health National Institute of Environmental Health Sciences
1:35 p.m.	**Exposure Science in the 21st Century: A Vision and a Strategy** (report released in 2012) Paul Gilman, Ph.D. Senior Vice President and Chief Sustainability Officer Covanta Energy Corporation
1:50 p.m.	**Advancing the Next Generation (NexGen) of Risk Assessment: Public Dialogue Conference** (meeting held in 2011) Ila Cote, Ph.D., DABT Senior Science Advisor National Center for Environmental Assessment Office of Research and Development U.S. Environmental Protection Agency
2:05 p.m.	**Committee on the Design and Evaluation of Safer Chemical Substitutions: A Framework to Inform Government and Industry Decisions** (status update) Marilee Shelton-Davenport, Ph.D. Senior Program Officer Division on Earth and Life Studies The National Academies
2:20 p.m.	Panel Discussion

3:05 p.m.   **Break** (20 minutes)

Session 4: Improved Approaches to Chemical Prioritization for Risk Assessment and Risk Management

Objectives:

- Build on the presentations in Session 2 that highlighted existing chemical prioritization frameworks from a variety of settings.
- Discuss better approaches to chemical prioritization (forward looking) to help inform targeted testing schemes and improve risk assessment and management strategies.

Moderator: Andrew Maguire, Ph.D., Roundtable Member

3:25 p.m.   **Bench-Level Scientific Innovation**

> Richard Judson, Ph.D.
> National Center for Computational Toxicology
> Office of Research and Development
> U.S. Environmental Protection Agency

3:40 p.m.   **State-Led Innovation: California**

> Gina Soloman, M.D., M.P.H.
> Deputy Secretary for Science and Health
> California Environmental Protection Agency

3:55 p.m.   **Country-Level Innovation: Government of Canada**

> Heather Patterson
> Assessment Strategies Division
> Existing Substances Risk Assessment Bureau
> Safe Environments Directorate
> Health Canada

4:10 p.m.   **Industry Innovation: American Chemistry Council**

> Christina Franz, J.D.
> Senior Director, Regulatory and Technical Affairs
> American Chemistry Council

4:25 p.m.   **Discussion**

5:30 p.m.   **Adjourn for the Day**

November 8, 2013

8:40 a.m.   **Welcome Back**

Lynn Goldman, M.D., M.P.H.
Roundtable Vice-Chair

Session 5: Actions to Reduce the Risk of Chemicals in Our Society

Objectives:

- Discuss the topic of prevention from an environmental stewardship approach.
- Cover broad concepts of sustainability and economics, and more technical issues of developing safer chemical alternatives.
- Provide industry examples of voluntary efforts to reduce use of chemicals in consumer products and internal decision making that led to these outcomes.

Moderator: Al McGartland, Ph.D., Director of the National Center for Environmental Economics, U.S. EPA

8:45 a.m.   **NIEHS Sustainability and Green Programs**

Trisha Castranio
Sustainability Analyst
National Institute of Environmental Health
Sciences

9:00 a.m.   **Johnson & Johnson Case Study**

Zephanie Jordan
Vice President Global Regulatory Affairs and
Product Stewardship
Johnson & Johnson

9:15 a.m.   **The Dow Chemical Company Case Study**

Connie Deford
Director, Product Sustainability and Compliance
The Dow Chemical Company

9:30 a.m.   **American Chemical Society Green Chemistry Institute® Case Study**

David J. C. Constable, Ph.D.
Director
American Chemical Society Green Chemistry
Institute®

9:45 a.m.	**SIN (Substitute It Now!) List and GreenScreen® Case Study**
	Beverley Thorpe Consulting Co-Director, Communications and Advocacy Clean Production Action
10:00 a.m.	**Massachusetts Toxics Use Reduction Institute (TURI) Case Study**
	Liz Harriman, M.S. Deputy Director, Toxic Use Reduction Institute University of Massachusetts, Lowell
10:15 a.m.	**Bullitt Center Case Study**
	Joseph David Sustainability Program Manager Point 32
10:30 a.m.	**Discussion**
11:15 a.m.	**Break** (15 minutes)

Session 6: Reflections on the Workshop and Concluding Remarks

Objective:

- Provide a moderated panel discussion to synthesize previous presentations and discussions (focusing on Sessions 1, 2, 4, and 5) and thoughts on possible next steps.

Moderator: Frank Loy, Roundtable Chair

11:30 a.m.	**Panel Discussion**
	Kimberly Thigpen Tart (synthesis of Session 1) Program Analyst, Office of Policy, Planning, and Evaluation National Institute of Environmental Health Sciences
	Bernard D. Goldstein, M.D. (synthesis of Session 2) Professor Emeritus of Environmental and Occupational Health University of Pittsburgh Graduate School of Public Health

Harold Zenick, Ph.D. (synthesis of Session 4)
Director, National Health and Environmental
Effects Research Laboratory
Office of Research and Development
U.S. Environmental Protection Agency

Lynn Goldman, M.D., M.P.H. (synthesis of
Session 5)
Roundtable Vice-Chair
Dean, School of Public Health and Health
Services
George Washington University

12:10 p.m.	**Discussion**
12:30 p.m.	**Adjourn**

C

Speaker Biosketches

John M. Balbus, M.D., M.P.H., serves as senior advisor for public health at the National Institute of Environmental Health Sciences (NIEHS). In this capacity he serves as the U.S. Department of Health and Human Services principal to the U.S. Global Change Research Program, for which he also co-chairs the Interagency Cross-Cutting Group on Climate Change and Human Health. Dr. Balbus' background combines training and experience in clinical medicine with expertise in epidemiology, toxicology, and risk sciences. He has authored studies and lectures on global climate change and health, transportation-related air pollution, the toxic effects of chemicals, and regulatory approaches to protecting susceptible subpopulations. Before joining NIEHS, Dr. Balbus was chief health scientist at the Environmental Defense Fund. He served on the faculty of The George Washington University, where he was founding director of the Center for Risk Science and Public Health, founding co-director of the Mid-Atlantic Center for Children's Health and the Environment, and acting chairman of the Department of Environmental and Occupational Health. He is a member of the Institute of Medicine's Roundtable on Environmental Health Sciences, Research, and Medicine.

Trisha Castranio holds the position of sustainability analyst for the National Institute of Environmental Health Sciences (NIEHS). Since NIEHS created the position in 2009, Ms. Castranio has been responsible for overseeing the institute's sustainable, eco-friendly, and green business practices. She develops sustainability policies and environmental management goals for NIEHS and is responsible for evaluating the effectiveness of NIEHS's stewardship initiatives (including promoting green chemistry and sustainable laboratories). Prior to her role as sustainability analyst, Ms. Castranio was a scientific researcher for NIEHS. In this role, she evaluated the role of environmental toxicants on early embryonic mouse development.

Wendy Cleland-Hamnett, J.D., is the director of the Office of Pollution Prevention and Toxics in the Office of Chemical Safety and Pollution Prevention at the U.S. Environmental Protection Agency (EPA). She has served in this position since 2009 and as the deputy of the Office several years before that. As the director, Ms. Cleland-Hamnett oversees the EPA's new and existing chemicals programs, numerous safer chemical and pollution prevention activities, enhanced efforts to make chemical information more accessible to the public, and a range of efforts to manage lead, formaldehyde, and other legacy chemicals. Ms. Cleland-Hamnett began her career at the EPA in 1979 and has worked across the agency in a number of offices and capacities, including the Office of Environmental Information, the Office of Policy, and the Administrator's Office.

David J. C. Constable, Ph.D., is the director of the American Chemical Society's Green Chemistry Institute®, which strives to catalyze and enable the implementation of green chemistry and engineering throughout the global chemical enterprise. From the end of September 2011 until January 2013, Dr. Constable worked as the owner and principal at Sustainability Foresights, LLC. The consultancy was directed toward assisting companies with sustainability, green chemistry, energy, environment, health, and safety programs. Prior to this, he was the corporate vice president of Energy, Environment, Safety, and Health (ESH) at Lockheed Martin. In that role, he led the ESH organization, provided leadership to improve corporate-wide performance in ESH, and guided the development of Lockheed Martin positions on emerging regulatory and legislative ESH issues. Prior to joining Lockheed Martin, Dr. Constable was the director of Operational Sustainability in the Corporate Environment, Health, and Safety Department at GlaxoSmithKline where he led the integration of sustainability, life-cycle inventory assessment, and green technology activities into existing business processes.

Ila Cote, Ph.D., DABT, is senior science advisor in the U.S. Environmental Protection Agency's (EPA's) National Center for Environmental Assessment. This Center conducts the EPA health assessments used to support Agency decision making. She is a board-certified toxicologist and has worked in the area of environmental risk assessment and science policy for the past 25 years. She is an adjunct professor at the University of Colorado Department of Molecular, Cellular, and Developmental Biology and a former faculty member of the New York University Medical Center's Department of Environmental Medicine.

Joseph David is the sustainability program manager at Point32, a Seattle real estate company focused on land use, development, and construction. In this role, he is leading Point32's efforts to secure the Living Building Challenge certification for the Bullitt Center, which is targeted to be the greenest commercial building yet constructed. He has spearheaded efforts to design a vetting process for meeting the Living Building Challenge's strict material "red list" exclusion requirements and has quickly become an industry leader in the field. Mr. David's prior design experience includes a performing arts theater, multi-unit affordable housing, and commercial core and shell projects. He has also worked in the renewable energy field, where he permitted and installed wind turbines throughout the Rocky Mountain west.

Connie Deford is the director of Global Product Sustainability and Compliance at the Dow Chemical Company. In this role, she is responsible for leading Dow's global product sustainability organization and program and for development and implementation of Dow's strategy on the Toxic Substances Control Act. This includes advocacy work within federal and state legislative and regulatory arenas and promoting Dow's leadership in chemical management as part of Dow's 2015 Sustainability Goals. Ms. Deford, a 29-year Dow employee, has held multiple positions within the company. Most recently, she was the director of Dow's Global Environmental Health and Safety (EH&S) Regulatory Organization where she oversaw development of Dow's European REACH Program. For the 15 years prior, she served as a Global EH&S product leader, where she had responsibility for product regulatory compliance, product stewardship, and providing EH&S input into development of business strategies.

Richard Denison, Ph.D., is a senior scientist at the Environmental Defense Fund with 27 years of experience in the environmental arena, specializing in policy, hazard and risk assessment, and management for industrial chemicals and nanomaterials. He has testified before various congressional committees on the need for fundamental reform of U.S. policy toward industrial chemicals and on nanomaterial safety research needs. Dr. Denison is a member of the National Academy of Sciences' Board on Environmental Studies and Toxicology and its Standing

Committee on Emerging Science for Environmental Health Decisions. He serves on the Green Ribbon Science Panel for California's Green Chemistry Initiative. He is a member of the National Academy of Sciences' Committee to Develop a Research Strategy for Environmental, Health, and Safety Aspects of Engineered Nanomaterials. In addition, he was a member the Environmental Defense Fund team that worked jointly with the DuPont Corporation to develop a framework governing responsible development, production, use, and disposal of nanoscale materials.

Christina Franz, J.D., is the senior director of regulatory and technical affairs at the American Chemistry Council (ACC). She is a policy advocate on health and chemical regulatory issues affecting the ACC member companies, concentrating in particular on the Toxic Substances Control Act, the High Production Volume Program, and product stewardship issues. She earned her J.D. from Loyola University Chicago School of Law and is admitted to the bars of the state of Illinois and the District of Columbia.

Paul Gilman, Ph.D., joined Covanta in 2008 as Covanta Energy's first senior vice president and chief sustainability officer. He is responsible for Covanta's safety, health and environmental compliance programs, corporate communications, and sustainability initiatives that further reduce Covanta's environmental impact while increasing the use of its technologies. Before joining Covanta, Dr. Gilman was the director of the Oak Ridge Center for Advanced Studies. He served as the assistant administrator for Research and Development and science advisor at the U.S. Environmental Protection Agency (EPA) from 2002 until 2004. Prior to joining the EPA, he was director for Policy Planning at Celera Genomics. He was previously the executive director of life sciences and agriculture divisions of the National Research Council of the National Academy of Sciences and Engineering. In addition, Mr. Gilman has held several senior government positions, including associate director of the White House Office of Management and Budget for Natural Resources, Energy, and Science, and executive assistant to the secretary of energy for technical matters.

Lynn R. Goldman, M.D., M.P.H., is an American public health physician, trained as a pediatrician and epidemiologist. Now Dean of the George Washington University School of Public Health, she is perhaps best known for her role in helping craft the Food Protection Act passed by Congress in 1996, the first national environmental law to explicitly require measures to protect children from pesticides. In 1993, Dr. Goldman was appointed by President Bill Clinton and confirmed by the U.S. Senate as Assistant Administrator for Toxic Substances at the U.S. Environmental Protection Agency (EPA), becoming the first physician to serve in this capacity. During her 5 years at the EPA, from 1993 to 1998, she promoted pesticide legislation reform, assessment of industrial-chemical hazards, and children's health issues. Dr. Goldman is vice chair of the Institute of Medicine's Roundtable on Environmental Health Sciences, Research, and Medicine

Bernard D. Goldstein, M.D., is emeritus professor of environmental and occupational health and former dean of the University of Pittsburgh Graduate School of Public Health. He is a physician, board certified in Internal Medicine, Hematology, and Toxicology. Dr. Goldstein is an elected member of the Institute of Medicine (IOM) of the National Academies and of the American Society for Clinical Investigation. His experience includes service as Assistant Administrator for Research and Development of the U.S. Environmental Protection Agency (EPA) from 1983 to 1985. In 2001 he came to the University of Pittsburgh from New Jersey where he had been the founding director of the Environmental and Occupational Health Sciences Institute, a joint program of Rutgers University and Robert Wood Johnson Medical School. He has chaired more than a dozen National Research Council and the IOM committees primarily related to environmental health issues. He has been president of the Society for Risk Analysis; and has chaired the National Institutes of Health Toxicology Study Section, the EPA's Clean Air Scientific Advisory Committee, the National Board of Public Health Examiners, and the Research Committee of the Health Effects Institute.

William E. Halperin, Dr.P.H., M.D., M.P.H., is chair and professor of the Department of Preventive Medicine and Community Health at the New Jersey Medical School and professor and associate dean of the Rutgers School of Public Health. Dr. Halperin worked as a physician and epidemiologist for the Centers for Disease Control and Prevention from 1975 to 2000. He began as an epidemic intelligence service officer, then directed a branch responsible for preplanned large-scale epidemiologic studies of occupational exposures, and later served as deputy director for the National Institute for Occupational Safety and Health. Dr. Halperin has authored more than 125 peer-reviewed scientific articles, and other chapters and books on various aspects of public health, including occupational disease and injury, public health surveillance, periodic medical surveillance, infectious disease epidemiology, child labor, clinical preventive medicine, and other areas. Dr. Halperin serves on the Board on Environmental Studies and Toxicology of the National Academy of Sciences.

Liz Harriman is the deputy director of the Toxics Use Reduction Institute at the University of Massachusetts, Lowell, and is responsible for managing the operations and research functions of the Institute. In her 14 years working at the Institute, she has provided technical research and support services to Massachusetts companies with the goal of identifying safer alternatives to toxic chemicals used in manufacturing and products. Her most recent work was the formation of industry supply chain workgroups to help companies comply with international regulations that restrict the use of certain chemicals, including lead and brominated flame retardants. Other technical work she has conducted includes design for the environment, chemical substitution, and analysis of industry progress in reducing waste and toxic chemical use in products.

Zephanie Jordan is vice president of Global Regulatory Affairs and Product Stewardship at Johnson & Johnson Family of Consumer Companies. In this role, she is responsible for leading the development of the global regulatory strategies for new products in the R&D pipeline across a broad portfolio of health and personal care products. She is also responsible for developing the regulatory policy agenda and leading the product stewardship initiatives and strategy as it pertains to ingredient policies as part of the broader sustainability framework. This work involves evaluating scientific, regulatory, and social trends and collaborating with stakeholders within and external to Johnson & Johnson across the globe. Ms. Jordan has worked in the consumer health care industry for more than 20 years and has been with Johnson & Johnson for almost 6 years primarily in regulatory and medical affairs leadership positions.

Richard Judson, Ph.D., works for the U.S. Environmental Protection Agency (EPA) National Center for Computational Toxicology, where he develops databases and computer applications to model and predict toxicological effects of a wide range of chemicals. He is a member of the

EPA ToxCast team where he leads the effort in bioinformatics. His team has developed the ACToR (Aggregated Computational Toxicology Resource) database and application which compiles all publicly available data on environmental chemicals. Dr. Judson has authored research publications in areas including computational biology and chemistry, bioinformatics, genomics, human genetics, toxicology, and applied mathematics. Prior to joining the EPA, Dr. Judson was founder of GAMA BioConsulting, a bioinformatics consulting company. From 1999 to 2006, Dr. Judson was Senior Vice President and Chief Scientific Officer with Genaissance Pharmaceuticals. Previously, he held research positions at CuraGen from 1997 to 1998 and Sandia National Laboratories from 1990 to 1996.

Frank Loy, LL.B., has served in the Department of State in four administrations. His portfolio included developing U.S. international policy and conducting negotiations in the fields of the environment and climate change, human rights, the promotion of democracy, refugees and humanitarian affairs, and counter-narcotics. In 2011 President Obama named him the U.S. Alternate Representative to the United Nations General Assembly. At present he serves on the boards of numerous nonprofit organizations. In the field of the environment these include Resources for the Future (former chair), Environmental Defense Fund (former chair), The Nature Conservancy, C2ES, and ecoAmerica (chair). He also chairs the boards of Population Services International and the Arthur Burns Fellowship Program and serves on the boards of the American Institute for Contemporary German Studies and The Washington Ballet. Mr. Loy is chair of the Institute of Medicine's Roundtable on Environmental Health Sciences, Research, and Medicine.

Canice Nolan, Ph.D., is the senior coordinator for global health for the European Commission Directorate General for Health and Consumers. In this capacity he is also the team leader for International Affairs in the Public Health Directorate. Dr. Nolan joined the European Commission in 1991 in the Directorate General for Research where he was responsible for program management on Environment, Health, and Chemical Safety. This work mainly focused on chemicals risk assessment methodologies, water quality, air pollution epidemiology, and endocrine disruption. In 1998 he moved within the Commission to the Directorate General for Agriculture where he was responsible for pesticides residues legislation. In 1999, following the reorganization of the Commission, this work moved to the Directorate General for Health and Consumers where he was head of the Plant Protection Products Sector—responsible for the management and development of legislation on pesticides evaluation, authorization, and use as well as the setting and monitoring of maximum residue limits for pesticides in food and feed. From 2004 to 2008, Dr. Nolan was head of the Health, Food Safety, and Consumer Affairs Section at the Delegation of the European Commission to the United States, based in Washington, DC.

Heather Patterson is a senior evaluator in the Healthy Environments and Consumer Safety Branch of Health Canada. Having worked on the assessment of existing substances under the Canadian Environmental Protection Act (CEPA) since 1999, she has been part of many science and policy initiatives to advance the assessment and management of industrial chemicals. She has worked directly on categorization of the Canadian Domestic Substances List and contributed to numerous assessments. Her current work involves developing innovative approaches for prioritization and assessment of chemicals, including methods for rapid screening and triaging of high or low concern chemicals.

Marilee Shelton-Davenport, Ph.D., is a senior program officer with the Board on Life Sciences at the National Research Council (NRC), where she has worked on a variety of biology projects since 1999. Presently, she is director of the Standing Committee on Use of Emerging Science for Environmental Health Decisions, which facilitates discussions on scientific advances for the identification, quantification, and control of environmental impacts on human health. She also directs a consensus committee on the Design and Evaluation of Safer Chemical Substitutions: A Framework to Inform Government and Industry Decisions. Before coming to the NRC, Dr. Shelton-Davenport worked at CNN as a mass media fellow of the American Association of the Advancement of Science.

Gina Solomon, M.D., M.P.H., was appointed by Governor Edmund G. Brown Jr. in April 2012 to serve as Deputy Secretary for Science and Health at the California Environmental Protection Agency. Prior to this, she was a senior scientist at the Natural Resources Defense Council since 1996 and has been on the faculty in the Division of Occupational and Environmental Medicine at the University of California, San Francisco (UCSF) since 1997, where she still holds the title of clinical professor of health sciences. Dr. Soloman has served on numerous scientific committees for the State of California, the U.S. Environmental Protection Agency, the National Toxicology Program, and the National Academy of Sciences. She is on the editorial board of the journal *Environmental Health Perspectives* and serves regularly as a peer reviewer for numerous scientific journals. Her prior work has included research on diesel exhaust and asthma, endocrine-disrupting chemicals, pesticides, environmental contaminants in New Orleans after Hurricane Katrina, the health implications of the 2010 Gulf oil spill, and the health effects of climate change. Dr. Soloman is board certified in both internal medicine and occupational and environmental medicine and is licensed to practice medicine in California.

Kimberly Thigpen Tart, J.D., is a program analyst in the Office of Program, Planning, and Evaluation at the National Institute of Environmental Health Sciences (NIEHS). Prior to this she served as news editor of the Institute's journal, *Environmental Health Perspectives*, for 15 years, and on detail to the Office of the Director under Dr. Sam Wilson. Her current focus areas include issues of climate change and human health, global environmental health, prevention research, and research policy and translation. She represents the NIEHS to the National Institutes of Health Prevention Research Coordinating Committee and the U.S. Global Change Research Program and as a member of the Interagency Climate Change and Human Health Working Group.

Beverley Thorpe is the consulting co-director of communications and advocacy at Clean Production Action. She has researched and promoted clean production strategies internationally since 1986 and was a co-founder of the International Programme on Cleaner Production at the United Nations Environment Programme. She was the first clean production liaison and technical expert for Greenpeace International on chemical and waste issues. During this time, she initiated the first English language campaigns against polyvinyl chloride (PVC) plastic and PVC waste and helped achieve a global ban on ocean incineration of hazardous wastes. Her current focus is the promotion of green chemistry within government policy and company practices and she continues to train, teach, and publish materials that advance clean production strategies internationally. Ms. Thorpe is a current board member of the Green Chemistry Network, the Story of Stuff, and Greenpeace Canada.

Michael P. Walls, J.D., is vice president of Regulatory and Technical Affairs of the American Chemistry Council (ACC). He has been with the

ACC for 22 years and has experience in a wide range of U.S. domestic chemical regulatory issues, including the Toxic Substances Control Act (TSCA), the Emergency Planning and Community Right to Know Act (EPCRA), and the Resource Conservation and Recovery Act (RCRA). His experience also includes work on international chemical regulatory issues, including the European Commission's regulation for Registration, Evaluation, Authorisation, and Restriction of Chemicals (REACH). Mr. Walls has represented the industry in several international chemical negotiations, and in support of U.S. ratification and implementation of those agreements. Mr. Walls began his work at the ACC in the Office of the General Counsel, where he provided legal advice on a range of international environmental and trade issues and product regulation. Before joining the ACC, Mr. Walls was in private law practice in Washington, DC, where he represented domestic chemical manufacturers.

Nsedu Obot Witherspoon, M.P.H., serves as the Executive Director for the Children's Environmental Health Network (CEHN), where her responsibilities include successfully organizing, leading, and managing policy, education and training, and science-related programs. She is a leader in the field of children's environmental health, serving on the Children's Health Protection Advisory Committee for the U.S. Environmental Protection Agency and is a member of the Institute of Medicine's Roundtable on Environmental Health Sciences, Research, and Medicine. Ms. Witherspoon is a past member of the National Association of Environmental Health Sciences Council and past coordinator of the National Institute for Environmental Health Sciences Public Interest Partners. She is a member of the Friends of the Columbia Center for Children's Environmental Health and a strategy advisor for the California Breast Cancer Prevention Initiatives project. Ms. Witherspoon is a long-standing leader in the Environment Section of the American Public Health Association and former executive board member for the Association.

Harold Zenick, Ph.D., is director of the National Health and Environmental Effects Research Laboratory in the Office of Research and Development in the U.S. Environmental Protection Agency (EPA). Before coming to the EPA, he spent 13 years in academia with the Department of Environmental Health in the University of Cincinnati Medical School, preceded by an appointment at New Mexico Highlands University. Dr. Zenick serves on the Executive Board to the National Toxicology Program and as the

EPA's liaison to the Board of Scientific Councilors for the National Center for Environmental Health/Agency for Toxic Substances and Disease Registry. He has participated on a number of prominent national and federal work groups and currently serves as co-chair of the Toxics and Risk Subcommittee under the auspices of the Committee on Environment, Natural Resources, and Sustainability in the Office of Science, Technology, and Policy. His current interests are in integrating human health and ecological risk assessment, strengthening the linkages between environmental and public health agendas and agencies, the promotion of sustainability and sustainable problem solutions as a critical consideration in the EPA's decision making, and the application of emerging computational, informational, and molecule sciences in improving toxicity testing and risk assessment practices.